WEST AFRICAN AGRICULTURE

M.F.
1932

NATIVE BLACKSMITHS WORKING ON PLOUGH (see p. 67).

West African Agriculture

BY

O. T. FAULKNER, C.M.G., B.A.

AND

J. R. MACKIE, M.C., B.Sc.
Nigerian Agricultural Department

CAMBRIDGE
AT THE UNIVERSITY PRESS
1933

CAMBRIDGE UNIVERSITY PRESS
Cambridge, New York, Melbourne, Madrid, Cape Town,
Singapore, São Paulo, Delhi, Mexico City

Cambridge University Press
The Edinburgh Building, Cambridge CB2 8RU, UK

Published in the United States of America by Cambridge University Press, New York

www.cambridge.org
Information on this title: www.cambridge.org/9781107625358

First published 1933
First paperback edition 2013

A catalogue record for this publication is available from the British Library

ISBN 978-1-107-62535-8 Paperback

CONTENTS

PREFACE

IT is hoped that the reader will not feel defrauded by the title of this book when he finds that it deals much more with agriculture in Nigeria than in the other colonies. The limitations of our first-hand knowledge of the latter is not the only reason for our concentrating so much on Nigeria. In the four countries there are roughly similar areas: for instance, the Northern Territories of the Gold Coast resemble the Northern Provinces of Nigeria. Even were we able to do so, we should be led into wearisome detail if we attempted to recount fully the differences in the agriculture of such corresponding belts. Yet it is hoped that a reliable account of a particular agricultural practice in Northern Nigeria may assist a resident of the Northern Territories of the Gold Coast to understand the corresponding practice there.

This book is written especially for the candidates for Government service in West Africa in the Administrative and Agricultural Departments; but we hope it may prove useful or interesting to those who intend to come out as missionaries. We have attempted to produce neither a standard text-book, nor a scientific treatise. Thus, matters such as climate and soil and the botany of the various crops have been very shortly discussed. Such information can be obtained from other sources. Our object has been to deal with some of the problems connected with farming in these countries and to emphasize the methods by which it is attempted to solve these problems. The economic side of the subject has been particularly stressed, as this is an aspect of tropical agriculture which has been but little dealt with in text-books.

The details of the experiments on which statements are based have commonly been omitted. They are to be found in the annual bulletins of the Agricultural Department of Nigeria, and some of the more important have also been collected together in a special bulletin on "The Maintenance of Fertility by Green Manuring".

The first part of the book is devoted to general subjects; the second, to some of the more important details about individual crops.

O. T. F.
J. R. M.

IBADAN
NIGERIA
1932

PART I
GENERAL

CHAPTER I

INTRODUCTORY

THE application of scientific methods of investigation to the problems of agriculture in West Africa is a comparatively recent development. The only agencies that can undertake this are the Governments through their Agricultural Departments, and in their present form these departments are really a post-War development. It was the activities of traders that first led to the establishment of Colonial Government in West Africa, and this fact had a considerable influence for many years on the general policy of the Governments. Yet neither the early Governments nor even the Chartered Niger Company, took the view that their only duty or interest was to protect the traders and facilitate their operations. At a very early date many valuable plants were introduced from other parts of the tropics in the hope of their being adopted by the natives of the country, and eventually Botanic Gardens were established and officers appointed, whose special function was this introduction of new crops and economic plants. But the main object in view was still an immediate increase in export trade, and this tendency persisted even when Agricultural Departments were established. Indeed the motive for their establishment seems to have been chiefly the hope of inducing the people, by more consistent propaganda, to adopt new crops—which they had frequently proved reluctant to do, in spite of persuasion and promise of future profit. "Quick returns" in the shape of increased production of export crops were similarly expected of the new Agricultural Departments, and their efforts to achieve these quick returns left them little opportunity for such a study of local farming and local conditions as alone could show what improvements were really feasible and likely to commend themselves to the native farmer. In any case the departments at first lacked adequate funds and staff. None the less they did much useful spade work; and by the time the War started, they had been able to define in some degree the problems before them; and to indicate the possibilities of progress. The staffs and funds of the departments were just being increased when the War caused all their activities to be suspended until a fresh start was made in 1921 or 1922.

The principle which guides all the actions of the Governments of West Africa to-day is the principle of trusteeship. In accordance with this principle it is the duty of the European Government, at all events of a heavily populated native dependency, to keep as its first and foremost object, not the benefit of European trade and the production of raw materials for the industries of Europe, but the moral and material advancement of the natives of the country. This is no new doctrine. It guided the actions of many administrators, district officers "in the bush", no less than Governors in the council chamber, long before it was officially enunciated. It is only in recent years, however, that this policy has been clearly defined and all its logical consequences universally accepted as the basic policy of the Governments of West Africa. Following this policy it becomes one of the duties of Government to study all the problems of local agriculture and the application of scientific methods to its improvement. Thus this policy led, immediately after the War, to a reorganization of the Agricultural Departments, and to their being supplied with the necessary funds, equipment and buildings. The purpose of the departments is now considered to be, not merely to stimulate production for export by native farmers, but to try to assist them to increased prosperity and well-being in all directions. In order to accomplish this it is necessary not merely to study what export crops can be grown, but also to try to increase the quantity or quality of the native foodstuffs. The economics of native agriculture thus become the foundation of agricultural policy. Incidentally it may be noted that this same policy is the one which in the long run will lead to the maximum production of raw materials, for if labour is saved in the production of foodstuffs, it becomes available for production for export. The "mandate" is a dual one, concerned with the material advancement of the people to take their part in the world's work, as well as with their cultural progress; and their material advancement is at present, at all events, bound up with the production of raw materials for export, after they have provided for their own food.

Instead of directly attempting to persuade the native farmer to grow certain crops which are required in trade, the agricultural officer now asks "Will this crop be profitable to the farmer? What assistance does he need in order to make it profitable for him? Is the variety of the crop which he grows well suited to local conditions? Is it the best that can be found? Does he need assistance in marketing its produce to good advantage? Can I ascertain for him by experiments whether a new crop would be suitable and profitable for him?" The agriculturalist in fact now tries to look at everything from the native farmer's point of view. It may be mentioned that the farmer's real views are by no means always the same as the facile statements which he will make to a casual enquirer, not really with intent to deceive, but rather because courtesy demands that he should give some answer to every question. The aim of the administrative officer is equally the prosperity of the native farmer and an appreciation of his problems. Co-operation between the agricultural and the administrative officers should thus be easy; and if this co-operation has not always been as close as it should be, the reason has been a lack of mutual appreciation of their common aim.

It is not surprising that the African farmer is suspicious of advice given by Europeans. His own methods have been evolved and adapted during many generations, so that they suit local conditions and also suit his economic position, his social arrangements, his psychology and his tastes. It will generally be found that by them he obtains a maximum return from a minimum of labour. True he often prefers a method that is slow rather than one that is quicker but involves harder work for a shorter time. This is due to the fact that he does the labour himself and therefore regards it from a different view point from one who pays for it by the day or hour. A European will often think that the methods are either inefficient or tedious and that he could easily suggest something better and more up-to-date. Unless the subject has been very thoroughly studied and every effort made to judge it from the native farmer's point of view, the suggestion is likely to be a misguided one. Provided it is strictly a suggestion, very tentatively put forward, perhaps no harm is done; but in the past Europeans have gone much further than merely to suggest, and have recommended, advised, persuaded, almost forced, the farmer to adopt their proposals, often without having first attempted to ascertain whether they were acceptable to him. Many instances could be quoted where this has been done and where the attempt was a complete failure. Such failures not only discredit the European in the eyes of the native farmer, but arouse in him a justifiable suspicion of all new ideas, which suspicion, once acquired, is not readily forgotten. Again, the prevalent idea that the native farmer is excessively conservative is largely due to the mistakes of Europeans in the past. These mistakes have been made by both agricultural and administrative officers. In the opinion of the authors, the native farmer in general is certainly not more conservative than the average English farmer; indeed many native farmers are much less conservative than most English farmers. If they can see that a new thing is worth trying they will willingly try it; but they very naturally resent being pressed to take up something in which they do not believe, or which they do not yet understand. The development of the production of cocoa in the Gold Coast, and of groundnuts in Northern Nigeria are examples of the way in which the native can develop a new industry almost entirely unaided, when he is satisfied that it is sound; and the failure of American cotton in Southern Nigeria is an example of how obstinate he can be when an attempt is made to force upon him something in which he does not believe.

While the statements that the native farmer knows his own business and that his methods are specially adapted to his own conditions are largely true, yet it does not follow that the European agricultural officer can do nothing to improve native farming. For one thing, the economic conditions to which the native system is adapted may be those of the past; and when changes in these conditions are involved there is an opportunity to help the farmer to adjust himself to them. Again, in evolving his methods he had not the advantage of scientific knowledge, nor the ability to carry out scientific experiments. The scientific study of tropical agriculture is in its infancy, and there is no doubt that some day there will be revolutionary changes in West African methods. But even while studying and preparing for these bigger changes, the agriculturalist can do a great deal to improve existing practices and crops; provided that, before suggesting anything to the native farmer, it is first proved by experiment to be in every way sound. This then is the new point of view from which the tropical agriculturalist has to judge everything connected with his work. Every idea must first be tested experimentally, and while conducting the experiments, the experimenter must attempt to see his idea from the farmer's point of view. This method may seem slow at first, but it is sure, and by this means alone is it possible to avoid the fatal mistake of losing the native farmer's confidence. At the same time, such work enables the experimenter to become an experienced local farmer himself.

The Agricultural Departments in West Africa are sometimes criticized on the grounds that in spite of the large sums of money spent on agricultural research, no very tangible results have as yet been produced, and native agriculture is in precisely the same state as it was ten or twelve years ago. The statement is not entirely true, and moreover it must be remembered that scientific investigation into agricultural problems is a slow process; for experiments, even with annual crops, have to be repeated over a series of years in order to take into account weather and other seasonal variations before a conclusion can safely be drawn. Experiments on the treatment of permanent crops naturally take still longer. Also, in West Africa, the investigator has a difficulty that he does not encounter in more advanced countries, in that he cannot take, as his starting point, the local general knowledge of the innumerable details of farming such as seed rates, and the best time for carrying out all operations; but he has to devote much effort to elucidating these elementary questions for himself before he can effectively study any more advanced matters. If we are to be sure that the suggestions that we lay before the native farmer are thoroughly sound, we cannot hurry the preliminary experimental work. Once we are able to see, from the success of the pioneers who have adopted them, that our suggestions are good, then rapid extension is feasible.

Since the test of all our results and conclusions is whether or not they are considered "sound" by the native farmer, it remains to consider what "soundness" means to him. In England, where land is scarce, rent and expenses high, the utility or otherwise of any agricultural improvement is judged largely by the profit or loss per acre, and anything which will increase the profit per acre is generally of merit, and likely to be adopted by the farmer. This, however, is not always the case in the tropics. Here, land is plentiful except in a few well-defined areas, such as the thickly populated area around Kano, in Nigeria, and parts of the cocoa belt of the Gold Coast. Generally no rent is paid, every native has a right to a piece of land, and among many tribes money, as such, beyond being a means of paying the tax, has still comparatively little meaning, so that small differences of profit and loss per acre are comparatively unimportant. The thing which really counts is labour. Everything has to be done by hand and the test of soundness is the return per normal day's work. A man and his family can only do so many days' work per year, and this strictly limits the amount of land which can be cultivated. Assuming that his farm is the maximum which a farmer and his family can manage if they work hard, then no change, however valuable it may be in cash, is an improvement if it necessitates the expenditure of much more labour. This is very clearly illustrated by the attempt that was made to introduce American cotton in Southern Nigeria. The crop grew well and the yields obtained from experimental plots were at least as good as those obtained from native cottons, while the produce fetched much more per pound than native seed cotton. Yet when seed of this cotton was offered to the native farmers they repeatedly showed—in deeds though not in words—that they would have nothing to do with it. It was subsequently seen that the point which had been completely missed was that of the return per day's work. American cotton would not do well in competition with other crops; it was only a success when grown as a sole crop. The planting of cotton as a sole crop entailed the labour of clearing more land for the purpose. The native therefore wisely preferred to grow his own less valuable native cotton, which could be grown on the same land with other crops, yielding, unlike the American cotton in these circumstances, not of course a full return, but a very fair one. When recently the farmer was offered an improved native cotton which he could interplant with his other crops in his old way, so that it entailed no extra expenditure of labour, he adopted it with remarkable celerity.

Like farmers in other parts of the world, the native farmer has his busy seasons and his slack seasons. The planting and harvesting seasons are always busy times, but there are periods between them which are comparatively slack. At these slack times he is prepared to regard a new idea more from the profit or loss per acre point of view; and he is prepared to do more work at these times if it will pay him to do so. Thus green manuring in Northern Nigeria was found to be unsuitable, because it meant preparing the land and sowing the beans at the busiest time of the year; but the making of hay for the dry-season feeding of cattle, if proved by further experience to be sound in other respects, is likely to be acceptable, since it can be done during one of the slack periods. This question of the spreading out of the available labour through the year is of paramount importance to any agriculturalist who is trying to devise better systems of agriculture by means of rotations.

Transport is another factor of great importance which has to be considered when estimating the soundness of any new suggestion. In a country where much transport is still done by carriers, the marketing of crops involves a serious drain on the labour available for farming, if the carrying has to be done during the busy season. In Northern Nigeria during the long dry season the people can market their crops without interference with work on the farm. This also applies to cocoa and cotton in the Southern Provinces. But apart from this, the native has a very shrewd idea of the value of carrier transport; it is a form of labour which he dislikes; and, in proportion to its value, there is a limit to the distance over which he is prepared to carry a head load of produce. Thus very few groundnuts are grown for export in areas remote from suitable forms of transport; but cotton is grown much farther back, because of its greater value per pound. The advent of a railway or motor road in Northern Nigeria therefore causes an increase in groundnut production, because then only can land which is unsuitable for cotton be profitably used for groundnuts. And although experiments might show that cotton will thrive and produce excellent yields in the remote parts of Bornu, yet it would be useless to try to induce the native to grow it in large quantities for export, since with existing transport facilities it would not pay him to do so. Similarly there are areas in the Southern Provinces of Nigeria where cocoa will grow, but where the cost of marketing is too high to make it profitable.

CHAPTER 2

CLIMATE AND SOIL

THE climatic feature which is common to all our West African colonies is the division of the year into two well-defined seasons, the wet and the dry. Since these colonies are all situated entirely within the tropics, they are of course subject to tropical temperatures; but temperature is influenced by local conditions such as altitude and nearness to the coast, whereas the division into two seasons is constant throughout the whole of West Africa. As a broad generalization one may say that the length of the wet season and the total rainfall decrease as one goes inland from the coast, and that at the same time the average temperature increases; the highest temperatures and the lowest rainfall are found in the extreme north of Nigeria.

The rainy season occurs during the summer months, from April to September, and during the rains the prevailing wind comes from the south-west and is therefore sometimes spoken of as the "South-west monsoon of the Coast". West Africa during the summer is situated within the South-east Trade Wind belt, but the Trade Winds are deflected from their course by the heating up of the Sahara and blow from the south-west instead of from the south-east, bringing with them the moisture-laden clouds from the ocean which cause the rainfall. After the end of September the rains cease somewhat abruptly and the wind veers round to the north or north-west. This northerly wind, known as the "Harmattan ", is in marked contrast to the south-west wind. It blows straight across the Sahara, and instead of being moist it is intensely dry and is laden with dust particles, which at times are so thick that visibility is reduced to a few hundred yards. As soon as the Harmattan sets in everything becomes desiccated and all plant growth practically ceases. Its effect at first resembles that of an early frost in England. Owing to its extreme dryness, and the amount of evaporation it causes, it is at first a cool wind and the temperature falls rapidly, especially at night; but later on, in February, March, and early April, it becomes hot and scorching; and it is during this period that the highest temperatures are recorded.

Near the coast the first rains occur in March and continue into November or December, so that the duration of the dry season is only about three or four months at most; but as one goes northwards this state of affairs gradually changes and in the extreme north the relative length of the season is reversed. At Sokoto for example there is very little rain until the end of May or early in June, and a storm after the middle of September is exceptional.

Although the Gambia is situated on the coast its climate resembles that of Northern Nigeria rather than that of the coast, in that the rains do not set in until June, reach a maximum in August, and end in October.

There is a narrow belt along the coast which has a climate of its own, being mainly influenced by land and sea breezes which are very local in their effect. Included in this belt is the area which stretches along the coast of the Gold Coast from Secondi to Accra. From its position on the coast it might be expected that this area

would have an average annual rainfall of about 80
inches, but it is in fact exceptionally dry, its rainfall
being about 25 inches per year. This peculiarity is
possibly due to the fact that the coast at this point
runs from south-west to north-east, and the prevailing
wind therefore blows parallel to it, instead of at right
angles.

Inland from this coastal belt, and including the
greater part of Sierra Leone, the Gold Coast and
Southern Nigeria is a belt in which the rainfall during
the wet season has two maxima. In this belt June, July
and September are the wettest months and there is a
short break in August. The comparatively dry spell in
August is of the utmost importance to the farmer, as it
enables him to harvest his early crops of maize and
groundnuts and to get his land ready for the late crops.
The drying out of the land then, and the warming effect
of the sunshine, stimulate a fresh production and accumulation of nitrogen in the soil which is available for
the late crops.

The transition between this and the third belt which
includes almost the whole of Northern Nigeria and the
Gambia is more or less gradual. As one proceeds farther
inland the August break becomes less and less well
defined until finally August becomes the wettest month
of the year.

The wettest areas in West Africa are to be found in
Sierra Leone and the south-east corner of Nigeria (which
includes the Cameroons). The excessive rainfall in these
areas is caused by the presence of the hills behind Freetown and the Cameroon mountains respectively. The
rainfall at Freetown averages 170 inches per year, while

that of Calabar is 130 inches per year and over 400 inches per year have been recorded in the Cameroons.

The Bauchi plateau, by reason of its altitude, also has an exceptional climate, being very much cooler than the rest of West Africa.

Throughout the whole of West Africa the rainfall mainly comes in heavy thunderstorms or tornados; a completely wet day is unusual, and occurs only in the middle of the rains. The storms become more frequent but less severe as the rainy season advances, and then, after the middle of August, again increase in severity but occur at increasingly long intervals.

From the appearance of the vegetation one would suppose that there was a very great difference between the rainfall at say Ibadan in Southern Nigeria and Zaria in Northern Nigeria, but this is not the case. The average annual rainfall at Ibadan is about 50 inches, while that of Zaria is about 44 inches. Yet there is a striking contrast between the tropical forests of Southern Nigeria and the orchard bush of the north. North of Zaria there is a very marked diminution in rainfall— for example, the average rainfall at Sokoto is only 27 inches—and the appearance of the vegetation again changes. With the exception of the baobab and the dorowa there are few big trees, and thorn bush replaces the orchard bush.

There is in Nigeria, as also in the Gold Coast, a very marked climatic boundary, and within a few miles the character of the vegetation changes, with startling sud-denness, from tropical forest, and such crops as cocoa, kola and the oil palm, to open grassland and orchard bush in which the shea butter tree is the most conspicuous feature. This boundary is a ridge of high land which runs across the country roughly parallel to the coast both in the Gold Coast and in Nigeria.

The sudden change in the character of the vegetation seems to be mainly due to the effect of the Harmattan. The north side of these ridges is exposed to the full force of the Harmattan, whereas the south side is sheltered; and, although there is only a comparatively slight difference in the total rainfall, the tropical plants cannot stand the extreme desiccation which the Harmattan causes.

During the rains there is very little difference in temperature or humidity between north and south; but in the dry weather the temperature in the north rises to between 100° and 110° by day, and frequently falls to 55° or even lower at night, while the relative humidity is extremely low. In the south the temperature even in the dry weather seldom exceeds 90°, nor is the relative humidity ever very low.

SOILS

The area of West Africa is so large that it is difficult to generalize on the subject of soils, as somewhere or other it is probably possible to find almost every kind of soil which exists, if one only considers the composition of soils in the mechanical sense. That is to say there are sandy soils and clay soils, and every graduation between these two extremes. The predominating colour is red, and largely owing to this fact, and the presence of concretions of iron oxide either in the form of rock-like outcrops or as a crust a short distance below the surface, the soil of the greater part of West Africa is loosely termed laterite. Recently the term laterite has been restricted to soils of definite composition, such as seem to occur where temperature, rainfall and humidity are all relatively high for the greater part of the year.

In the south, where the rainfall is high and water is nearly always percolating through the soil, the fine particles tend to be washed into the subsoil, and the subsoil is often finer and contains more mineral plant food than the surface soil. Farther north, where the rainy season is shorter, the top soil contains more of the fine particles. Still farther north, along the edge of the Sahara, the soil is almost pure sand for a depth of several feet.

Differences in the texture of the soil are quite clearly understood by the native farmer, and he knows exactly which crops are suited to any particular type of soil. Where the population is thick he has very little opportunity to discriminate, and his food crops at any rate have to be grown on any land which is available, irrespective of quality; but where the land is plentiful he uses real discrimination. In Northern Nigeria, for example, other things being equal, groundnuts are grown on light sandy soil and cotton on heavy land.

The more difficult, and probably more important, soil problems in tropical countries are concerned with chemical composition of the soil, soil acidity, and the biological changes which take place within it. The investigation of these soil problems is in its infancy as far as the tropics are concerned, and our present ideas are largely based upon the results of research carried out in temperate countries. Frequently such results are not truly applicable to tropical soils.

Practically all soils in West Africa are deficient in

organic matter, and judged by the standards of temperate countries also generally appear to be very deficient
in calcium and phosphates. But in spite of such de-
ficiencies they are often extremely productive. It is
evident that the dynamics of the matter, i.e. the rate at
which plant food in the soil becomes available, are so
different in the tropics and in cooler climates, that the
static criteria by which soils in temperate zones are
judged are inapplicable to soils in the tropics.

The response of most tropical soils to even the smallest
dressing of nitrogen in an organic manure is truly remarkable, while their frequently poor response to
dressings of phosphatic manures is equally strange, and
still requires investigation.

Occasionally one does come across an area where
some deficiency or abnormality is so marked that the
soil is rendered definitely infertile. The Benin sands
and the soils of the Niger Delta are a case in point.
Apparently they are so acid and deficient in available
mineral plant food that many crops will hardly grow on
them at all. Yet some plants, e.g. the oil palm, grow
reasonably well upon them; and it will be interesting to
see if other crops will do equally well if the acidity is
reduced by the addition of lime. Most tropical soils
have a slightly acid reaction, but normally this is not
sufficient to act as a limiting factor to plant growth. The
usual tables for the lime requirement of soils are hardly
applicable to tropical soils, but the Delta soils mentioned
above are so extremely acid that it would seem that lime
should improve them greatly.

The chemical composition of the soil, particularly
with reference to its phosphate and calcium content,

has recently become important in connection with the efforts that are now being made in West Africa to improve the local breeds of live stock. It is a well-established fact that mineral salts have an important function in the metabolism of animals. If they are very deficient in the diet, pathological conditions in the animal will develop, or at least its rate of growth and capacity for production will be strictly limited. It has already been established that the ordinary bush grasses are deficient in phosphate and calcium, and the relation between the mineral contents of plants and that of the soil on which they are growing is now being closely studied, with a view to improving the diet of the live stock.

The biological changes which take place in the soil are also the subject of careful investigations, as they are concerned with the decomposition of organic matter in the soil, which has an important bearing on such questions as green manuring and the effect on the soil of farmyard manure.

It will be seen therefore that our present knowledge of tropical soils is as yet very limited, and that the subject is bristling with problems which are not by any means purely theoretical. They have indeed a very important application to the farm of every individual peasant. If, for example, the work of the soil chemist should lead to the extension of cocoa growing in parts of Southern Nigeria where it cannot be grown at present or to a general all-round improvement in the live stock of Northern Nigeria, he will have added greatly to the prosperity of the people, and the money spent on his investigations will be amply repaid.

CHAPTER 3

POLITICAL ECONOMY

THERE are practically no European planters in British West Africa, and relatively very few large estates under native ownership. Apart from the fact that the climate hardly permits of Europeans making a permanent home in the country, it has always been the policy of Government to keep the land for the people and to prohibit the alienation of any large blocks of land to any individual, African or foreign. The land is regarded as belonging to the local community, which, as a community, has absolute security of tenure and most of the rights of a freeholder.

The rights of the individual farmer within the community are determined by local "native law and custom". This is generally unwritten and frequently exceedingly difficult to translate into terms that are readily intelligible to a European. Local native law and custom, moreover, naturally varies vastly from place to place; and it is also liable at times to very rapid changes in response to altered economic conditions. Since the law and custom is unwritten, the native courts which administer it may revise it, or may retard its changes. Experience shows, however, that the courts cannot permanently and entirely prevent the custom changing in response to changing conditions. Nor indeed can they attain complete consistency in their interpretation of the customs. Consequently it is extremely difficult to generalize about the rights of individuals. But it may be said that generally the individual farmer (or family) has a right, with some qualifications, to continue permanently to farm the land which he uses regularly. But apparently it is also a principle of native law and custom that every adult married male of a village, or whatever the communal unit may be, has an equal right to the use of a piece of land. If, therefore, the population of a village is already large enough to use all its land, and the population of married males increases by one, some readjustment is necessary, and has to be carried out by the village head or the village council of elders.

These two principles do not conflict, and present no difficulty, so long as land is plentiful. They seem to be preserved even when land is scarce, so long as the people are truly primitive; but they must inevitably fail, and have in fact failed, before the inroads of civilization. They have broken down, not only before the new Western, but equally before the several old African civilizations. The individual is unwilling to give up a part of his land whenever the number of adult married males increases. And moreover, in a highly organized state, there are inevitably some individuals who are powerful enough to assert individual claims and to maintain them against the community. Such claims, though they may at first be established on no other principle than that of " might is right ", eventually acquire public recognition; and gradually it comes about that even less powerful individuals are considered to have some permanent individual rights. Thus among the more advanced communities the individual holding of a family (in the narrow European sense) is not really regarded as liable to reduction as the community increases. Fortunately, the more advanced a community becomes, the more capable are the individuals of migrating when they are overcrowded; and fortunately, although in all the British West African colonies there are districts where there is no land to spare, there are still, in all of them, districts where there is plenty of fertile land unoccupied; and other difficulties in the way of migration, though still great, are gradually decreasing. It is not very inaccurate to say that any native of British West Africa who so wishes can have a piece of land to cultivate, sufficient for his livelihood.

As a consequence of the local systems of tenure which have been described, or as a result of their former existence even in places where they have since been greatly modified, the great bulk of the people of British West Africa are peasants. Each has the use of as much land as a man and his family can use, and they have at least considerable security of individual permanent occupancy of that land. Even where, as in some places in Nigeria, the whole village periodically moves its cultivation around a series of blocks which it uses in rotation, each individual is said to return each time to his own plot within the block; though it would puzzle a stranger to understand how the boundary marks are recognized when the land has been unused for years and is covered by thick bush.

In those parts where, before the Pax Britannica, a great slave raid or a tribal war was almost an annual dry-season event, as among the Hausas and Yorubas in Nigeria, the people naturally adopted the habit of living in towns or large villages. They either walked out daily to the farms, or frequently maintained two houses and lived on the farm more or less constantly only during the busy rainy season, which was always a season of peace. But elsewhere, as for instance among the Munshis and Ibos in Nigeria, the native social and political organization was less highly developed, and large armies were rarely collected and then too slowly to effect surprise. In such districts the people have always lived in small villages near their farms, and there were no towns. In either case the former habits are gradually changing in response to modern requirements, and while the farmers live near their farms, there are towns which form centres of trade and industry.

The West African peasant naturally has little capital. The assets of a well-to-do peasant consist of his wives and children, his simple hut and household utensils, one or two cutlasses and hoes, a few poultry and two or three head of sheep or goats. The stock live in the compound and feed in the neighbouring bush, and no attempt is made to fold them on the land or to improve their quality either for meat or milk. The milk of animals is not used by the people of Southern Nigeria even for children; to drink the milk of animals is to them an unnatural and repulsive idea.

In Northern Nigeria there is a growing class of settled farmers who own a few cattle, but in the main the cattle industry is in the hands of the Fulani herdsmen who are not farmers at all. They are entirely nomadic and usually have no farm or settled place of abode, but wander through the bush with their herds, which graze by day and whenever possible are "kraaled" on the cultivated lands at night. These herdsmen live mostly on milk and meat, while the peasant farmer lives chiefly on grain. In spite of some interchange, the diet of each class is deficient in the products of the other.

In the south, yams and maize are the staple food crops; but the northern farmer lives chiefly on millet ("gero", *Pennisetum*) and guinea corn ("dawa", *Sorghum*). Meat is very scarce indeed in the south and, in order to obtain animal food, some tribes will eat rats, dogs, and even snakes. In the north, meat, although not a staple article of the daily diet of the ordinary farmer, is usually obtainable; and is eaten occasionally by everyone with a frequency proportionate to his means, for it is the favourite luxury. Around each compound, whether in the north or the south, there is usually a small patch of land which is permanently cultivated, and is manured by the refuse from the compound. It is usually planted with small vegetables, and plants that supply minor domestic needs, fibres for rope, dyes such as henna or indigo, gourds, and often tobacco.

The main crops are grown on the farm proper which may be some distance from the compound. The native farmer does not ordinarily farm the whole of his available land. Every year, part of it is under cultivation and part of it is under bush conditions. When the cultivated land shows signs of exhaustion, it is allowed to revert to bush again and a new piece is cleared. (This system is described more fully in the chapter on "Shifting cultivation".) The area actually cultivated annually by each individual farmer varies according to the amount of labour which he has available, but in general it is about three acres. In the north there is an occasional bigger farmer, who farms on quite a considerable scale, but he is an exception. While the area actually under cultivation is thus comparatively small, that over which each individual has rights is more considerable, and almost all the land in the more thickly populated areas is claimed by someone, whether it has ever been farmed or not. In the more thinly populated districts, rights to the occupancy of very extensive areas are often asserted by individuals, and still more often by village communities. But when these areas vastly exceed the amount of land which they could possibly farm, or to which they could even have enforced their claim in the days when might was right, it is arguable that the rights are only those of the collection of forest products and of hunting.

The work of cultivation is shared by all members of a family, and in most tribes the various operations are very definitely classified into men's work and women's work. The men generally do the heavy work such as clearing and making the original heaps of soil or ridges for planting; they then do certain definite major weedings and assist with the harder parts of harvesting, such as cutting and laying guinea corn stalks. The women, as a rule, do the planting, keep the land clean between the main weedings, harvest and market the crops. There are, however, great variations from tribe to tribe in these arrangements; generally the more primitive the tribe the heavier the work done by the women, whereas among the stricter Muhammedans the women do practically no farm work. Children are required to assist in many of these operations, but their special function is to keep the birds from the grain crops while the grain is ripening.

Forest trees which give valuable fruits, such as oil palms, are sometimes owned individually and sometimes communally. Individual ownership of oil palms

arises generally from the fact that the owner's ancestor
had planted the tree, or allowed it to grow up, when he
was farming the land on which it now stands. It frequently happens that the right to farm the land in
question has since passed to another, so that the man
who owns the tree so long as it is alive would have no
right in the land when it dies or if he cut it down. When
oil palms are communally owned, days for harvesting
are fixed by the elders, when each member of the community can harvest as much as he likes; or alternately
the whole area to be harvested is divided up, for the
occasion, between the harvesters. Shea nut trees are
generally regarded as entirely common property and the
collection and the preparation of the nuts and butter is
women's work.

All these arrangements have largely broken down
wherever cocoa has been planted. Any communal right
to the use of the land on which an individual has been
allowed to plant cocoa either ceases to exist or becomes
highly theoretical. Then, again, the wealth which cocoa
brings leads to the women being freed from their obli-
gations for heavy work in the fields. But the social
traditions and arrangements that have been described
still hold good, quite unshaken by "progress", in many
parts of all the West African colonies; and even in cocoa
belts the present arrangements are based largely upon
the old ones from which they have evolved.

Under the old conditions the native farmer is quite
independent of world trade. There are no landless people
and no unemployed. Only a serious shortage of rain or
a plague of locusts can cause a lack of food; and these
events occur but rarely even in the north, and never in

the coastal belts. The farmer can prosper without producing anything for export, or even for local sale, except so far as is necessary to provide his tax money. If he grows a surplus of money crops for export, the money so obtained is used to provide him with such luxuries as meat, cloth or clothes, tinned food, kerosene, or corrugated iron for his house. It is therefore mostly returned to the trading firms again almost at once. The village artisans—weavers, smiths, potters, and so on—are generally farmers as well, and in any event exchange their products directly with local farmers.

It might be expected that such a state of affairs would break down in the vicinity of large towns. As yet, however, it is only the large ports, such as Accra and Lagos, that have greatly influenced the agriculture of their neighbourhood. The large inland towns, although gradually changing, are still, so far as economics go, in effect large villages—very large sometimes, but still rather villages than towns, in that the great bulk of their inhabitants are farmers, and the town is fed by its own people farming in the immediate neighbourhood.

It is in the cocoa belt that the old economic conditions have changed most, and the change has proceeded much farther in the Gold Coast than in Nigeria. Many of the cocoa farmers of the Gold Coast have practically ceased to grow any food, and even the work of the cocoa fields is done by paid labourers. Food is obtained from other parts of the country or actually imported from abroad. This is naturally true also of the big ports and, to a less degree, of the mining areas. Sierra Leone, also, is largely dependent on imported rice, which is paid for by exports, chiefly of palm kernels. The local production of rice is, however, rapidly increasing, and it would seem that in time of real necessity it would not be very long before the people could greatly increase the supply of other native foods.

The degree to which the West African native generally is, or can be, self-contained, is a factor of the greatest importance to the economics of West Africa; for if the prices of export products are low, the great mass of the people merely have to give up "imported luxuries", in proportion to the fall of prices. They do not starve and there is very little unemployment or distress. This would seem to place the country in an extremely stable and strong position. The non-producers and the organizations dependent on trade—the trading firms, the railway, the motor transport workers—though adversely affected by the decrease of imports which follows a decrease in the *value* of exports, can still continue their operations so long as the *volume* of exports is maintained. The position of Government revenue is very similar: the portion derived from import duties inevitably falls if the value of exports decreases; but the direct taxation, which represents the major portion of the revenue, is not so directly dependent on the value of external trade. Now although the goods bought with the proceeds of the exports of produce are "luxuries", the producers are human, and therefore want their luxuries, and may be expected to continue production in spite of a considerable fall in the value of produce. For since they do not generally pay wages, they are not in such a position that, if their business does not pay, it had better be closed down rather than run at a loss when the price of the product falls to a certain level.

The exact limit at which the Nigerian farmer would decide that production is not worth while is a question that can hardly be answered. So far (1931) it may be said that the recent period of depression shows that the volume of exports is still fully maintained even at a relatively very low level of prices. In this period of considerable depression, West Africa has been affected to a degree which is relatively small compared to the difficulties which have beset other countries whose economics are based on the payment of wages.

As already pointed out, the situation of the cocoa belt of the Gold Coast is rather different from that of the rest of West Africa. What are luxuries in other parts are there necessities, and internal trade between the different parts of the country can no longer supply all needs. Thus a serious fall in the international price of cocoa affects the producers more acutely than it does in Nigeria. But, even so, the relatively small native cocoa planters of the Gold Coast have been able to meet bad times with less difficulty than large plantations could do; for although Gold Coast planters depend on paid labour the labourers are mostly men who have farms of their own on which they work during most of the year to grow their own food, and they only work on the cocoa farms in the dry weather to earn their luxuries, just as the Nigerian peasant produces for export; and so, like the peasant, they can when necessary work for less pay more readily than labourers who are entirely dependent on their wages. The Gold Coast cocoa planter has thus been better able to adjust himself to low prices than a large planting corporation, which must either make a profit or else eventually close down altogether.

The deliberate permanent maintenance of the system of peasant proprietorship in its entirety, as it is still maintained in Nigeria, is liable to several legitimate criticisms. One of these is based on the fact that the motive for the production for export is to obtain luxuries. Therefore it seems to follow that if the price of exports continues falling, a point must eventually be reached where the producer decides that "the game is not worth the candle", and gives up producing for export altogether until the price recovers. There is then no function for traders, no work for railways, no source of revenue for Government. Whereas, on the other hand, a wealthy planting company, when the price of its product falls, does not in fact cease operations as soon as it is no longer able to make ends meet, but will go on producing at a loss for a considerable time, because the loss so incurred, if not too long continued, is much less than the capital loss incurred by neglect of the estate. Indeed it has been shown by experience that such companies will go to great extremes to keep going in bad times, so as to retain their labour and organization in being, ready to take advantage of better times when they come. Only a small man, with no such elaborate organization, can afford to suspend operations temporarily. At present, discussion as to which will give in first, the West African peasant or the planting company, is still purely speculative, and so far experience has not provided the answer.

Digressing a little, there is another subject connected with bad times in West Africa that may be mentioned here. While it is indeed still true to say, as a broad generalization, that the West African native grows the necessities of his life and imports the luxuries, yet this statement must not be regarded as entirely accurate. Luxuries are by no means always imported; many "luxuries" are already almost necessities; and some of the export products are the subject of internal as well as external trade. The people of the south-east of Nigeria, who ten years ago wore a minimum of clothes but are now comparatively well clad with imported cloth, will not revert to nudity if they can possibly avoid it; and they may avoid it in a great degree by selling palm oil, which cannot be exported if its price is too low, to other parts of Nigeria where cotton goods are manufactured locally and palm oil is a luxury. Similarly the coastal belts in both Nigeria and the Gold Coast need meat from the interior and the interior needs palm oil from the coastal belt. Thus there is an internal commerce, which though it might not have arisen in the first instance had there been no export and import trades, will continue and may well even expand when the external trade is depressed.

Another criticism that is sometimes made of the West African peasants, especially those of Nigeria, is that when there is an opportunity for a suitable new planting industry they are extremely slow in taking advantage of it. The rate of expansion of cocoa production by the Nigerian peasant, for instance, is unfavourably compared with that of the rubber industry in "Planters" countries. This argument seems to be entirely fallacious. When the West African peasant seems to have been slow, it will be found on closer investigation to be due, not to characteristics which are inherent in his position as a peasant, but to difficulties that arise from the details, rather than the principles of the local system of land tenure; or from purely technical difficulties. Where neither of these operates against him, the West African peasant can develop an industry very quickly, as is shown by the very rapid growth of cocoa production in the Gold Coast and of the groundnut industry of Northern Nigeria.

It has also been pointed out by his critics that there are many tropical crops which a peasant cannot possibly produce because an elaborate factory is needed to prepare the product for export. Industries in this category are sugar, sisal, canned fruits and tobacco. As the peasant is unable to take up any of these industries, it is said that West Africa is unduly restricted to a few staple products. To this it may be answered that although it is undoubtedly unfortunate for any country to produce only one agricultural product, a vast variety is not necessary; and in point of fact, other tropical countries, which are not subject to the limitations imposed by peasant proprietorship, have generally concentrated on one or two products just as completely as West Africa does.

The most effective criticism that can be levied on economic grounds against the exclusion of planting companies is the fact that there are many tropical crops which fall into an intermediate category. These are products which the peasant could grow, but of which, under the conditions of modern trade, either the cultivation, the preparation, or the collection, grading, packing and marketing, involve so much technique that the peasant cannot succeed unless he is given a lead by European planters. The classical example of this is the rubber industry. Even with the example of the estates before them, the natives of the east were very slow to take up this new industry, because of its technical nature. Yet recently they have shown, very practically, that, given a lead, they can produce rubber effectively and cheaply. As products which fall into this category, or upon one or other of its borders, may be mentioned coffee, limes, fresh fruit for export, copra (coconuts), cinchona, indigo, tobacco, and even sugar. In most of these instances the native grower would not only have difficulty in learning the technique needed in the field, but would need the assistance of a central grading and packing house, or of a central factory with machinery.

On economic grounds alone, it must be admitted that this argument is hardly answerable. The natives of West Africa could produce many of these things probably more cheaply than the inhabitants of any other tropical country if they could be given the necessary lead and assistance, and it cannot be questioned that planters could give that lead more efficiently than the only possible alternative agency, namely, the Government. For one thing, since the Government does not depend for its livelihood on the continuance of any particular experiment of this kind, it is very liable not to continue it long enough for success. Thus a relatively large-scale experiment in coffee growing in Nigeria was given up by the Government after a few years because it was not paying; whereas it was subsequently seen that the price was then only temporarily depressed. The price rose rapidly a few weeks later, and if the plantation had been continued for one year more it would have paid handsomely, and in all probability would have given rise eventually to a valuable native industry. Again it is evident that if the Government of Malaya or Ceylon had established rubber plantations solely as an example for natives of the country, these estates would have been shut down as having been proved by experience to be unsuccessful ten or twenty years too soon. For it has taken the natives of those countries many years to learn the wisdom of following the example of the commercial plantations.

But while it can hardly be doubted that the presence of some planters in West Africa would be effective in this way, there are other considerations which carry the question outside the domain of agriculture.

CHAPTER 4

THE NATIVE FARM

I N such a vast area as West Africa, with such varying
climates as those described in Chapter 2, it is not
surprising that agriculture should also vary greatly
from place to place.

In the country near the coast, with its heavy rainfall
and short dry season, the life of the native farmer is
one long fight against the rank growth of vegetation.
Clearing new land covered with thick bush or forest is
a laborious operation; and, when the land is cleared, an
enormous amount of labour is continually required to
prevent the weeds and undergrowth from encroaching
on the cleared area and destroying the crop. The farms
therefore tend to be small and intensively cropped, and
are indeed more in the nature of gardens than farms.
Mixed cropping is the general practice, and the spacing
of the plants is as close as possible. Two crops of cereals
can be grown every year; and other crops can be interplanted with the yam crop, the most important crop of
this area. By such heavy cropping, a whole family can
live without difficulty on the produce of a comparatively
small piece of land. The following quite typical example
will give some idea of the intensity of the cropping on
native food farms in the vicinity of Ibadan, in Southern
Nigeria:

1st year. Bush cleared and burnt in July and late maize
    planted on the flat in September. Hills for yams made
    in November and yams planted in the same month.

2nd year. Early maize planted through the yams in March, cotton planted through the yams and maize in early August, maize harvested late in August. In addition, edible beans and gourds also grown in the cotton.

3rd year. Early maize with cassava planted through it.

4th year. Cassava.

5th year. Reverted to bush.

In the whole cycle of four years the land only received one thorough cultivation, namely when the hills were made for the yam crop. The labour of deep cultivation is reduced to a minimum, but weeding and shallow cultivation are done frequently. It may be mentioned here that experiments on the Government experimental farms in Southern Nigeria, while they have not yet indicated exactly for how long it is possible to continue to raise good crops without any deep cultivation, have shown that annual deep cultivation is not essential, nor does it give the increase of crop that might be expected.

As one proceeds inland from the coast there is a gradual change from this intensive gardening type of agriculture to an extensive type of farming in the far north. Between these two types there is a middle belt with an intermediate type of farming. The northern farmer experiences a long dry season, during which the ground is bare, and all farm work is necessarily at a standstill. He has only a short rainy season, varying from five to three months according to latitude, in which to grow his crops. His great aim therefore is to grow as much as possible during this short period, and the quantity of food or money crops which he can grow depends chiefly on the amount of land which he can cultivate. At the end of the dry season there is very little vegetation on the land, so that clearing is not the laborious operation that it is in the south; and afterwards weeds are a very much less serious factor than in a humid climate. Owing to the shortness of the season there is no possibility of getting two crops in the same year. Mixed cropping is generally practised only with the two cereals, millet (gero) and guinea corn. These crops are usually grown on the same land; but cotton, groundnuts, rice, beniseed, etc., are all as a rule grown as sole crops. A few rows of gero may be grown through groundnuts; and gourds and edible beans are often sparsely sown in the cotton and guinea corn fields respectively; but, in the sense in which it is practised by the southern farmer, mixed cropping is not seen in the north. Further, the northern farmer, unlike the southerner, has always to keep in mind the possibility of his crops failing from drought, and he has therefore to allow for a margin of safety. The failure of the early rains means a poor millet crop and the early onset of the Harmattan can seriously affect the guinea corn and cotton. It is many years since there was any general or complete failure of crops in Northern Nigeria, but there occurs periodically a season like that of 1926, when the shortage is serious enough to make food very dear in the following year.

The northern farmer usually has a fairly large area around his compound which is cropped every year, and this is lightly manured by the refuse from the compound or by the droppings of cattle which are herded on the land at night during the dry season. But apart from this, like the southern farmer, he practises shifting cultivation and depends upon the natural regenerative power of uncropped land to recover its fertility.

The northern farmer has to start work immediately a sufficiently heavy storm of rain falls in April to enable him to plant millet (gero). In order to save time, this is planted without any previous preparation of the land, and as soon as the land can be dug guinea corn must be planted. From this time onwards the weeding of these crops keeps him fully occupied until cotton has to be planted in July or early August, by which time the gero is ready for harvest. In the coastal country, although naturally there are the right times for planting the various crops, the long rainy season and uniform temperature save the farmer from the continual fight against time that is inevitable in less favoured regions.

There is another and wider difference between the farming of the south and that of the north, in that the climatic conditions of the south are suitable for the growth of permanent crops, such as cocoa, coffee, kola and the oil palm. While therefore the northern farmer has to depend entirely upon annual crops for both food and money, the southern farmer is, in many instances, able to depend upon permanent crops for his money and need only grow annual crops for food. The result of this is that while the southern farmer tends to become affluent and to do less and less manual work, the northern farmer, under existing conditions, must cultivate a larger area for a smaller return. On the other hand, local conditions, and the state of semi-feudalism which still exists in the large Emirates, have given rise, in the north, to a class of large landowner. These large farmers are able to command supplies of domestic labour which are not available to the ordinary peasant farmer, in the shape of freed slaves, and their descendants. The land of these big farmers is generally divided up into farms of the usual size, each managed by one "servant" or tenant. The master commonly interferes very little in the management of the farms and simply takes the crop at harvest time, leaving an amount considered sufficient for the servant's food for the year. In addition, he accepts a general responsibility for the servant, supplies him with clothing, and meets the expenses of such events as marriages and funerals in his family. The survival of this system of management, almost as inefficient a system as could well be conceived, is due largely to the inertia and servile temperament of the servants, who as yet have little desire for greater independence; but it is evident that it will sooner or later be replaced by some more business-like arrangement, to the benefit of both parties. In those parts of the coastal belt, e.g. in the Western Provinces of Nigeria and in parts of the Gold Coast, where the primitive West African village communism had been replaced in part by feudal arrangements, these latter were often rather those of villeinage than domestic slavery, and both have already largely broken down into those of landlord and tenant, or of independent farmer and labourer. These changes have especially accompanied the establishment of permanent crops.

It seems hardly necessary to draw attention to the great attractiveness of permanent crops in those regions where they can be grown. They need no special preparation of the land. The land is usually prepared for annual crops and the permanent crops are interplanted in them at suitable distances. Annual crops continue to be grown until the shade from the permanent ones becomes too thick.

It might be thought that by interplanting with annual crops in the early stages the young trees would receive a very poor start, but experiments have shown that, with oil palms at any rate, the young trees make better growth when interplanted with annual crops than under any other system. Once the permanent crop is well established little work is necessary. Very few weeds can grow under the canopy, for instance, of a cocoa plantation, especially when the trees are planted as closely as is customary in West Africa; so that practically the only labour is that of the harvesting and preparation of the crop. The time may come when these plantations will require more careful management, such as pruning and thinning; but this is apparently not essential at present and, once the trees come into bearing, the produce is almost all clear profit. The farmer is then able to retire to the town, build himself a superior house, live upon imported luxuries, and increase the number of his wives.

Thus it might seem that the straight short road to wealth and prosperity is to plant permanent crops wherever possible; and that the obvious duty of agricultural officers is to advocate this measure. There are, however, great dangers in such a course. For one thing, there is the danger that results from a country becoming too dependent on the international price of one product; for unfortunately the conditions in any one country or district generally make one permanent crop much more attractive than any other, so that everyone grows the same crop. Again there is always the possibility of widespread damage or destruction by some pest or disease, with results that are calamitous if everyone has concentrated on the one crop. Again the question of the planting of permanent crops is closely interwoven with that of land tenure, and the official adviser must be careful that he is not advising a farmer to establish a plantation on land to which he has no individual right. It is felt therefore that while it is justifiable to assist the would-be planter with technical instructions, or seeds or seedlings, and to help him to market his produce, it is dangerous to advise him to plant exclusively any one particular crop.

On the other hand, it is of little use for a Government to try to discourage the planting of any crop by propaganda, or to refuse to assist either in the obtaining of seed, or in the marketing of any particular crop, in order to avoid the dangers of too much dependence on that one product. If one crop is more profitable than others, the farmer will not be deterred from planting it by any exhortations not to do so. If the Government really wished to hinder its extension, the only effective way would be to make its cultivation less profitable through such measures as preferential taxation or customs duties, or by giving assistance or special privileges to anyone who plants other crops.

CHAPTER 5

SHIFTING CULTIVATION

THE practice known as "shifting cultivation" is almost universal throughout the whole of West Africa. Under this system the farmer clears a piece of land, crops it intensively for three or four years and then allows it to revert to bush again until it has regained its fertility, meanwhile clearing another area of bush land in order to make a new farm. The resting period may be anything from one year upwards, depending on the density of the population and consequent demand for land; but the commonest period is about four or five years. The continuance of such a system depends therefore upon abundance of land, and hitherto this condition has existed generally in West Africa. But there are large areas in Nigeria where it no longer exists and, with an increasing population and an expanding demand for the produce of tropical countries, it can only be a question of time before a shortage of land becomes more general.

The native does not deliberately destroy trees whose fruits are of economic value, but, when land is cleared for farming, such trees, especially young oil palms, receive a check to their growth, which is often quite deliberately given by heavy scorching and by the cutting off of many leaves. Even the stumps of valueless trees and large bushes are not destroyed; but their regrowth is checked temporarily by regular lopping so long as the land is in cultivation. The stumps of the bushy undergrowth have a great value in ensuring that, when land in due course reverts to fallow, it shall be a fallow of secondary bush rather than one of grass and annuals; for the former type seems to lead to a greater recovery of fertility, and also has the advantage that when the land is again cleared it is not infested with weeds.

Where there is high forest still available, it is preferred for cultivation to any secondary bush, so that shifting cultivation in such areas entails the continual destruction of forest, which is valuable both for its timber and for its effect on climate. In these circumstances shifting cultivation is obviously most wasteful. In large parts of Nigeria, especially in the provinces of the south-west, which have a moderate density of population, say 100 or 200 to the square mile, all high forest has already disappeared; and the shifting cultivation there would be more accurately described as a system of rotational "bush fallows", in which the time in fallow exceeds the time that the land is cultivated. There the system involves no deliberate waste and gives fair returns. If it were the most profitable one possible, it would not be objectionable, though it is clear that a permanent system would allow of more selection of the best land for farming, or leave more land for permanent crops or for forest reserves. But where, as in the Delta Provinces in Nigeria, especially Onitsha and Owerri, the population already numbers 400 or 500 and even more per square mile, the land has to be under cultivation for more than half the time. There the system has really broken down already, and the yields are very poor, even allowing for the inherently poor nature of the soil:

the farmers continue the system because they (and ourselves too in that area) know of nothing better. In the Gold Coast cocoa belt a somewhat similar position appears to be approaching, owing to so much of the land having been devoted to cocoa.

Shifting cultivation has no doubt been a primitive system in all countries, and in all countries has eventually been replaced as better methods have been evolved. The native farmer in West Africa realizes at least one objection to the system; for the labour of repeatedly clearing new land for cultivation is irksome; nor is he unaware that bigger crops can be grown by the use of manure, and everywhere he uses manure if it is available. In the north, cattle manure is highly valued, but, as has already been explained, most of the cattle belong to nomadic herdsmen; and although farmers pay to have the cattle herded on their land whenever possible, the demand greatly exceeds the supply. In the south, cattle cannot be kept, but small stock—sheep and goats—are kept, and their manure is used together with ashes and sweepings. These, however, do not go very far. In some parts, leaves and leaf-mould are actually carried in from the bush on to the farms; but in very densely populated areas this is hardly feasible, and in any case it is too laborious a process to be carried out all over a farm.

The problem of replacing shifting cultivation by permanent is therefore one of finding a supply of manure. But it must be noted that any manure, or any new system of farming to provide for manuring, must be more profitable than shifting cultivation. However primitive the old system may be, or however objectionable, because it involves the destruction of forest or the use of an excessive area of land, the farmer cannot be expected to give it up in favour of a new system which yields a less return for the same amount of labour. On our experimental farms in Nigeria the utmost attention has therefore been devoted to studying the economics of this process. In addition to calculations based on the labour of clearing and on the labour of growing crops, and the value of the returns, there is a very valuable criterion which is available by which to judge the soundness of any system which is studied. This criterion is "Can we by the new system farm at a profit, using daily paid labour?" The labourer in West Africa has to be paid a higher wage than the farmer earns on his own land. The West African, like any other peasant, would rather be independent and work some land for himself if he can earn as much in that way as he can as a labourer. Land is available for all, and men only become labourers instead of farmers because they earn more as labourers. If, therefore, a system of farming can be made to pay when carried out by labourers who receive a daily wage, that system must be economically sound. Therefore, in studying farming in Nigeria, most careful accounts of costs are kept and are considered on the bases both of profit per acre and of return per man-day.

It may be mentioned that Europeans frequently seem to attach an exaggerated importance to crop rotation, and to think that by simply rotating the crops the native farmer can avoid the necessity of shifting cultivation. This is not true: a series of crops grown regularly in rotation will eventually exhaust the soil just as surely as the growing of the same crop year after year unless the rotation in some way provides for a supply of "manure" in some form or other. Moreover, as already shown, the native farmer does already practise the rotation of crops.

The use of "artificial" manures has been studied to some extent in Nigeria by means of field experiments on a considerable scale. It might be expected, in view of the high cost of purchase and transport of such manures, and of the low scale of profit and loss on the farms, that it would be hardly possible for such manures to be profitable. And this is found to be generally the case when actual experiments are conducted; at the best, the return from the use of artificial manures has barely covered the outlay. This, of course, is not enough; no one wants to spend money, and take the risk of loss, merely in the hope of getting his money back again; and very much more profitable results than this would be needed before the native farmer, who has very little money at all, would spend it upon artificial manures. Experiments with such manures are still being continued, because of the possibility that they may be more profitable in certain special circumstances; but it is clear that they can offer no general solution of the problem of permanent cultivation.

From what has already been said, it will be realized that in considering the replacement of shifting cultivation by manuring, there are in Nigeria three quite separate sets of conditions to be considered. There is first the north, where generally land is not scarce and where cattle can be kept and their manure used on the farms, though the supply is still inadequate. Secondly, there are the parts of the coastal belts where the population is not too dense to allow of the land being allowed

to lie fallow for a sufficient proportion of time to maintain its fertility. Then there is the third case, chiefly
represented by the Delta Provinces, where the soil is
very poor, the rainfall very heavy, and the population
so dense that land can only be allowed to lie fallow for
quite inadequate periods. In the first two cases it is
primarily a question of replacing systems which are not
entirely unsatisfactory, but yet should be susceptible of
improvement, by a method that would give a greater
return for the same amount of labour; and the question
largely turns on whether manuring is more profitable
than rotational "bush fallowing" with its concomitant
repeated breaking of "new" land. In the third instance,
the situation is different; it is a case of finding some new
system which will give good crops in the place of the
very poor ones that are obtained at present, and which
will be profitable and within the means of the people.

For the first two sets of conditions it is considered
that systems have been found which are both feasible
and much more profitable than the existing system, in
the form of "mixed farming" in the Northern Provinces
and "green manuring" in the South-western Provinces.
Each of these forms the subject of a later chapter. But
on the poor and over-populated soils of the Delta, it
must be admitted that no solution has yet been found.
Those soils are of a very peculiar character, and one of
their peculiarities is a high acidity, which is occasionally
so extreme that some crops, including those used as
green manures, will not grow at all, or, if they can be
grown, seem to be much less efficacious in restoring
fertility to the land than they are on normal soils. We
have not there, by green manuring, been able to grow

as good crops as can. be obtained after a "bush fallow" of a few years' duration. The acidity and other peculiarities of this soil are now being closely studied, in the hope that some other line of attack may be found; such, for instance, as a combination of liming and green manuring.

CHAPTER 6

GREEN MANURING

IN a district of heavy rainfall, the moment bush is cleared and land is left bare fertility is lost at a surprisingly rapid rate. Washing and leaching are the principal factors which contribute to this loss, and in any system of permanent cultivation the necessity for keeping the ground covered with vegetation, and taking every possible step to prevent wash, is of paramount importance. Europeans have been slow to realize the importance of this point, and there is no doubt that the planters of Ceylon and Malaya are suffering for it to-day. The native farmer, by his method of mixed cropping and by leaving the stumps of the bush in the ground, effectively minimizes the loss by leaching, but his system of growing crops on hills instead of ridges, as is common in many parts of West Africa, allows much wash to occur, with the natural consequence of a loss of fine soil.

Growing plants have a remarkable influence in causing rain to enter the soil, instead of allowing it to run off from the surface, as any one can see for himself by merely comparing a lawn and a piece of bare ground during a heavy shower. But this is by no means all. Except when a soil is exceedingly dry, natural processes are always going on by which plant food is converted into the soluble state. Especially the nitrogen contained in the insoluble decayed organic matter is always rapidly changing into a soluble form. Unless there are plants growing on the land to take it up as fast as it becomes soluble, every rain leaches out much nitrogen. Experiments prove that the weight of crop which can be grown in West Africa, as commonly elsewhere, depends primarily on the supply of nitrogen. Bush fallows restore fertility by the accumulation of nitrogen in insoluble organic matter, which, when the land is again cultivated. will decompose and provide soluble nitrogen again There are also recuperative bacterial processes, by which enough nitrogen from the air is accumulated to provide for the growth of bush or grasses; but under cultivation the rate of loss by leaching, let alone the absorption by the crops, is much more rapid than the recuperative processes. In a tropical climate, under a heavy annual rainfall, a piece of land which is cultivated and uncropped very soon becomes practically sterile, instead of increasing in fertility as it will do in a drier climate.

The mineral plant foods are not so rapidly lost, and in a tropical soil appear generally to pass into an available condition at a rate which is ample to provide for plant growth. In field experiments we have had very little return from the application of phosphate or potash, except when the rate of growth of the crop has simultaneously been greatly stimulated by the use of soluble nitrogenous manure. On account of the presence of soft phosphate deposits in Nigeria, and the consequent possibility of supplying ground "rock" phosphate at a comparatively low cost, this manure has received especial attention; but the results obtained have been unsatisfactory. The increases in yield resulting from the applications were small, and they were not greatly increased even when the material was ground as finely as possible. It should be noted here that these remarks apply to the gravelly soils of Southern Nigeria; on the clay soils at Zaria the results were rather more promising, especially when the phosphate was applied to leguminous crops or when combined with farmyard manure.

The great need of the lateritic soils of Southern Nigeria is for nitrogen. And apart from the economic considerations already mentioned in the discussion of artificial manures in the last chapter, the rapid loss of nitrates that occurs under a heavy rainfall renders it essential that the nitrogen should be applied in the form of organic manures. Any organic form of nitrogen gives excellent results, and even such a small dressing as $\frac{1}{2}$ ton of cotton seed per acre was found to result in a very large increase in yield, and also had a considerable residual effect over the following two or three years.

Green manuring is the process of growing a crop simply for the sake of turning it all into the soil as a manure. The plant grown for this purpose is usually one of those belonging to the leguminous family—the family which includes beans, peas and clovers. The plants of this family have the unique property of collecting nitrogen directly from the air through the agency of bacteria which live in nodules on their roots. As has already been explained, exhausted land will increase in fertility if anything is grown on it—even bush or grass—and allowed to return to the soil, since by so doing the nitrogen which is continually being fixed by certain soil bacteria is preserved and accumulated instead of being washed away, and can be used when the soil is again cultivated. Practical experiments have, however, shown that the process is greatly accelerated if leguminous plants are used. For instance, at Ilorin in the south-west of Nigeria, field experiments have shown that "*Mucuna*" (*Styzolobium aterrimum*; the Bengal bean), grown for a few months and then dug in, causes a much greater increase in the succeeding crop of maize than a crop of grass grown in the same way. The beneficial effect of clover was appreciated by the Romans, and the value of various "pulses" (small beans or peas) in soil improvement was utilized in India and China long before that. Actual green manuring in which the whole plant, and not merely the roots or residue, is returned to the soil has long been practised to a limited extent in India, and has comparatively recently been used in Europe as one of the chief means of improving very poor soils. Still more recently, leguminous "cover crops" have been widely used in Ceylon, Malaya and the Dutch East Indies to prevent losses by the washing and leaching of the soil of tea and rubber estates. (It should be explained that "green manure" and "cover crop" are practically synonymous terms; the former is used when the manuring effect of the crop is the chief consideration, and the latter when its protective value is of primary importance.)

But although leguminous crops have thus been widely used in various special circumstances, or to augment other forms of manuring, it seems that, previous to the recent experiments in Nigeria, no investigation had ever been carried out to ascertain whether it is possible and economically feasible to maintain land under annual cropping at a high state of fertility by continually growing green manure crops in regular rotation with others. Continuous experiments carried out at Ibadan and Ilorin in Nigeria, on these lines, during the last ten years, have produced results which have exceeded all expectations and have shown conclusively that, by the use of green manure, ordinary land can be kept under cultivation for at least this length of time without any decrease in fertility. The crops are now heavier than at first, weeds have been kept down, and the system pays.

Since green manures were thus being used in a way which had not been tried before, it was necessary to study many aspects of the subject about which little or no information was available. The results have caused us to hold very different ideas about the effects and functions of green manures from those which one would gather from a study of the technical literature of the subject. Since we need practical and economic results, our study of the subject has been conducted by field experiments, in which the manure is grown and treated in various ways on neighbouring plots and the result simply judged by the weight of the succeeding crops—a method which contrasts with that more commonly followed, where attention is largely concentrated on decomposition of the green manure and the resultant changes in the composition of the soil. Consequently the facts which we have ascertained are unquestionably true economically and practically, in contrast to some of the deductions that have been made from the other type of study; on the other hand, while we have ascertained the end-result in terms of effect on succeeding crops, there is still room for much work to be done to establish with certainty the way in which this end-result is produced.

One of the first questions that arose was would the

soil become "bean sick"? It is well known in England that clover and field beans cannot be too frequently grown on the same land without it becoming "clover sick" or "bean sick"; and that, when this has occurred, the land needs several years' rest from that crop before it will recover. It was therefore necessary to test many varieties of bean in order to find out if sickness would result from the use of all or any particular varieties. The varieties most exhaustively tested were cowpeas, *Canavalia*, *Dolichos* and *Mucuna*. Cowpeas and *Canavalia* were soon found to be in every way less suitable than either *Dolichos* or *Mucuna*, but there was at first very little to choose between these two. *Mucuna* produced rather more growth than *Dolichos* and appeared to last longer during the dry weather, but *Dolichos* appeared to have a rather better effect on the following crops. However, after a few years, it was found that the land became "sick" from *Dolichos* and the crop was always severely affected by disease, so that in the end it had to be discarded and *Mucuna* is now used exclusively. The possibility, however, of finding something still better is not being neglected. It is extremely fortunate that a bean has been found which can be grown year after year on the same land without any ill effects, as the whole success of the system depends upon this. There are, of course, other leguminous plants which can be successfully used as green manures. *Crotalaria*, *Centrosema* and *Calapagonium* have all been tried and all have their uses. *Calapagonium* is particularly useful for growing under heavy shade and *Centrosema* and *Crotalaria* perhaps grow better than *Mucuna* on the acid soils of the Benin district; but for most Southern Nigerian conditions

Mucuna is almost ideal. It is easy to establish, and although it usually requires one weeding soon after it is planted, its growth thereafter is so rapid that it completely smothers weeds of every sort. By covering the ground rapidly and completely, it effectively prevents wash.

A question of considerable importance is whether the removal of the seed greatly lessens the value of the leguminous crop in improving the soil. This is a point on which experiments are still proceeding. But there seems to be no tropical bean which gives a heavy crop of seed: 200 or 300 lb. per acre being a very good yield. Now a good crop of field beans in England may yield ten times as much, and moreover the straw is removed from the field: yet the beneficial effect of beans on a succeeding wheat crop is very marked. It would therefore seem that the removal of the small quantity of seed that is yielded by tropical beans, without the removal of the straw, could do little to lessen the effect of the leguminous crop on those which succeed it. The obvious next step would therefore seem to be to use a bean of which the seed was valuable for export or local consumption. But here a great difficulty is found. All of the many tropical beans and pulses that yield good edible seed seem to produce slow-growing, poor, straggling plants; and many of them produce few seeds unless the plants are supported on sticks. They therefore have to be frequently weeded. The yields are hardly heavy enough to pay for the weeding; and, moreover, our experiments show that crops of such plants often have little effect in increasing the succeeding grain crop. On the other hand, the strong-growing beans, which keep the land clean and are effective green manures, such as *Mucuna*, *Dolichos* and *Amberique*, are inedible or barely edible. There seems to be no tropical bean or pulse which combines the two characters of palatability and rapid, bushy growth; nor, indeed, even one that effects a fair compromise between these two desirable characteristics. Yet there seems no reason why this should be the case, and, since we have failed to obtain such a bean by importing varieties from other countries, or by selecting them locally, efforts are now being made to produce one by cross-breeding.

The general idea commonly held in regard to green manuring seems to be that its benefit largely results from the increase of nitrogen and organic matter which comes from burying into the soil the aerial parts of the plants in the green state. The name indeed implies this, and a number of chemical investigations have been carried out, as a result of which statements are made as to the exact stage in the life of the plant when it is best to dig it in because the nitrogen content of the aerial parts is at a maximum; and, from similar reasoning, instructions are also given that the crop should be dug into a damp soil, immediately after cutting, without allowing it to dry, lest decomposition be delayed. A series of experiments have been carried out at Ibadan on the same land for a series of years by growing a green manure crop in the latter part of the rains, for the benefit of the maize crop that is grown in the succeeding season. These experiments show that it makes very little difference whether the green manure is cut at the end of the rains and buried green immediately, cut and allowed to decompose gradually before burial, or allowed to grow until it dies of drought before being dug. It is even found that there is little or no loss in the succeeding maize crop if the green manure is burnt *in situ*. The only thing which seems to make any substantial difference to the result obtained from green manuring is the complete removal from the field of the whole tops of the crop. Now, if the crop is burnt, the organic matter in the aerial parts, with the nitrogen which it contains, is lost in the burning; so the explanation of these facts would seem to be that the nitrogen which the soil gains in the process of green manuring, and which is of most importance, must be chiefly formed in the roots of the crop or in the soil itself, rather than in the aerial part of the plant. Even the burning of the tops is apparently satisfactory, presumably because the bulk of the nitrogen gained is untouched and the mineral matter in the tops is returned to the soil. The complete removal of the tops alone apparently results in decrease of the effect; this is due, presumably, to the loss of valuable mineral matter.

The crops obtained at Ibadan now, after ten years of rotational green manuring and cropping, are as heavy or even heavier than were obtained from newly cleared bush; and they are heavy and very quick-growing crops by any standard. It seemed probable, when the experiments were first started, that we should eventually reach a condition in which mineral food would become available to the plant at a rate which would be too slow to support such rapid growth. This has not been found to be the case, presumably because of the efficiency of the green manure crops in accumulating readily available mineral food as well as nitrogen.

Again, green manuring has generally been regarded as a process by which fertility is gradually accumulated, and crops which occupy the soil for a long time have often been used or recommended for this purpose. From our general experience, however, and from some actual experiments on the point, this seems to be wrong, at all events in the tropics. A short, quick-growing green manure, provided that it grows to a real crop, will apparently give almost, if not quite, as great an effect as one which occupies the soil much longer. We have even tried growing two green manure crops in immediate succession without appreciably increasing the effect. It would seem that, when this latter process is adopted, or when a long-period crop is used as green manure, one gets rather a cycle of changes at a certain level, than a steadily rising level of fertility, or at least the further improvement becomes exceedingly slow.

On the other hand, green manuring is generally regarded as being in the nature of a permanent improvement of the soil, the benefit of which will be seen for many years. Our experience is exactly the contrary; it seems to be rather of the nature of a quick-acting manure, of which the full benefit is obtained, and the effect fully exhausted, in at most two succeeding crops.

All these statements of course refer only to the tropical conditions and light soils of Nigeria, and may not be true in other circumstances. But they seem to be true here, and it is evident that a permanent system of farming based on green manuring will require that manures are grown with considerable frequency, but need not each occupy the soil for any long period.

The systems practised at Ibadan may be illustrated by one of the main rotations followed there:

1. Yams and cover crop.
2. Green manure and cotton.
3. Groundnuts and cover crop.
4. Maize and cover crop.

It will be seen that in this rotation there is one leguminous crop every year, and that the land is always kept covered, except for a short period between the sowing of the crop and the time that it actually covers the ground. After maize and yams even so much "gap" does not occur, for the cover crop is sown through the main crop some weeks before it is harvested. The respective parts played in the maintenance of fertility by the actual digging in of the green manure, and by the prevention of leaching through always having a crop on the ground, have not yet been fully ascertained; but keeping the ground covered is one of the essentials in farming in the tropics.

The frequency with which a green manure crop must be dug in if fertility is to be maintained has not yet been finally decided. In the example given above there is either an early or late green manure every season; and after ten years the crops are very heavy and rather better than at the beginning. Other experiments in which an early green manure is followed by maize or cotton every year, and *vice versa*, have confirmed the fact that one green manure or cover crop each year is ample; and we are inclined to think that this frequency is a little greater than is necessary. It is probable that one early green manure every three years, with a late cover crop somewhere in between, is ample to maintain fertility, provided that the land is never allowed to suffer from wash or leaching.

When these experiments were started it was customary to give the land a thorough cultivation for each crop, which in some cases meant two cultivations every year; and it was thought that the excellence of the yields obtained were perhaps due as much to the thorough cultivation as to the effect of green manuring. The rotation mentioned above, in spite of this expensive, frequent, deep cultivation, has proved profitable; the yields are much heavier than the native farmer obtains even from new land, and they pay for the cost of the cultivation and growing the green manures. But if this rotation is compared with the native rotation mentioned in Chapter 4, it will be seen that the frequency of deep cultivation has been very greatly increased. From this point of view, the system is one of big outlay for big returns instead of small outlay for small returns, and would therefore represent a very great change of method to the native farmer. Other experiments have, however, shown that such frequent deep cultivation is superfluous; and, provided that the land has a thorough cultivation every three years or so, there is nothing to be gained by more frequent deep cultivations. One deep cultivation every few years is exactly what the native farmer actually does, and the establishment of this point will do much to make the system more attractive to him. Rotations and methods are now being devised, and are being tested, in which the green manuring is included, but the deep cultivations are much less frequent.

Our object, however, is not so much to devise one

particular rotation to be presented to the native farmer as a cut and dried, approved system, as to be able to instruct him on more general lines, and to tell him what we have found that one can or cannot do, when land is maintained in high fertility; as for instance that, after an early maize crop, one can grow a good crop of cotton but not a good late maize crop; whilst on the other hand, one can, on land in good heart, grow a good early maize crop after a late one.

Elasticity must be a characteristic of any system which it is hoped will be taken up by the native farmer. It can hardly be expected that he will jump at a rigid and ready-made system and follow it in its entirety. If he is to adopt green manuring he must be allowed, and assisted, to adapt it to his own methods; to modify his own system rather than to abandon it.

One of the minor difficulties of the system is the question of seed supply. If the whole green manure crop is dug in while it is still green, as is necessary with an early green manure, it means that a special area must be set aside to supply seed. Here again, however, the native farmer may be expected to solve the problem himself. Such a question is apt to assume great importance on a large plantation where the quantity of seed required is very considerable. The native farm is so small that sufficient seed can be obtained from a very few plants set aside for the purpose.

It is generally considered in India that green manuring is not a feasible and profitable practice in a very dry climate without irrigation; for, even if a good green manure crop can be grown in these conditions, the yield of subsequent crops is too much limited by the rainfall to make the process economically sound. It has been found at Zaria in Northern Nigeria that a sufficiently heavy green manure crop can be grown there, and that digging in a crop of *Mucuna* leads to a very great increase in the following two crops of guinea corn; and it is just possible that the process is economically sound even at Zaria. But it is one that the farmer there could hardly be expected to take up. The reason is that at Zaria it is only possible to grow one crop a year instead of two, as in the south. For this reason, and owing to the short time available for cultivation, to grow a green manure crop practically means giving up an equal area of guinea corn which could have been grown instead. Thus the process is only profitable if the increase in succeeding grain crops more than compensates for the crop that was lost. In the experiment mentioned above, the return just exceeded the loss: but, as with artificial manures, something much better than this would be needed before any farmer could be expected to take it up. Experiments also have been made in the Northern Provinces to test the effect on succeeding crops of the very light crop of green manure that can be grown before cotton, which is the crop planted last; but such green manure crops were found to have no effect at all. It is evident that green manuring is not a practical proposition in this country except where the green manure crop and another crop can be grown in the same season. Fortunately, where the climate is too dry for that to be feasible, it is possible to keep cattle, and there is thus no need for any other system of manuring.

CHAPTER 7

MIXED FARMING

ALTHOUGH mixed farming is a Nigerian subject rather than a West African one at present, there will clearly eventually be an opportunity for this kind of farming in the Northern Territories of the Gold Coast and in the Gambia.

It has already been explained that permanent cultivation exists in Northern Nigeria around the big towns and villages wherever manure is available; that the ordinary farming is of the extensive type; that further extension is limited by the supply of labour; and that the country supports a vast number of cattle, which, however, are mainly owned by the pastoral Fulani herdsmen and not by the settled farmers. All the materials for building up a system of mixed farming are therefore already in existence.

In speaking of mixed farming, the picture which we have in mind is one in which every farmer owns cattle of his own, say two bullocks and one or two cows, together with his usual head of sheep, goats and fowls. He would keep his cattle in a pen and supply them with bedding, thus making farmyard manure of the highest quality all the year round. His bullocks would be used for ploughing, thus solving the labour problem; and his cows would breed calves and supply him with milk for his family or for sale. The calves would appreciate in value as the male animals come on for draught purposes and the females develop into milking cows. Such a system, if it became universal, could not fail vastly to increase the prosperity both of the individual and of the country. Indeed, it might well be asked why it has not become the custom of the country long ago.

Before the British rule, the settled farmer was commonly little more than a villein or slave; and, even if any farmer had the money to invest in cattle, he dared not do so, as they were apt to be forcibly taken away from him; or, at the best, such an obvious sign of increased wealth made him liable for excessive taxation. This state of affairs ended when the country was taken over by the British; and after a few years a number of settled farmers began to invest their savings in cattle. Recently the number of those who have done so has been steadily increasing. But, until only two or three years ago, the keeping of cattle still involved one risk which was so great that very few could afford to take it. That was the risk of an epidemic of rinderpest. Rinderpest is always present in Northern Nigeria, but in those years when it assumed epidemic proportions three-quarters of the cattle used to be completely wiped out, and a man's whole herd might be destroyed. So long as cattle were subject to such epidemics every few years, mixed farming was hardly feasible.

The recent success of the Veterinary Department in working out a system of preventative inoculations has altered the whole position. A farmer can have his cattle immunized on the payment of a very small sum; and the fact that most of the nomadic herds are already immunized greatly lessens the likelihood of a widespread epidemic, even among those animals that have not yet been treated.

It has been realized from the start that mixed farming and the introduction of ploughing should be treated as one and the same problem. The advantages of ploughing by means of bullocks, as compared with hand cultivation, are so obvious that there is little doubt that if this alone had been advocated, and progress measured in terms of the number of ploughs in use, extension would have been very rapid. But it was thought that to increase the area under cultivation without any manuring would be a mistake. For, in spite of shifting cultivation, the yields per acre are already low; and with the permanent cultivation of larger areas by ploughing they would soon have become poorer still, unless some provisions were made for increased manuring. It was therefore decided not to attempt to introduce ploughing by itself; but to wait until a complete system of mixed farming had been worked out and then to try to introduce the system as a whole. That this was the right method seems already to have been proved by experience; and although the details of a system of mixed farming have taken several years of experiment, so that apparent progress has been slow, we are confident that the system finally evolved is absolutely sound from the native farmer's point of view. It has been proved, by the experience of the small but growing band of farmers who have adopted the system, that the native farmer who now farms about three acres can, by keeping a pair of bullocks and using a plough, increase his cultivated area to ten or twelve acres; and, what is equally important, he can make enough manure to obtain better yields by manuring than he previously got by shifting cultivation. Thus the final effect is, that although his area has been extended, the farming becomes more intensive. That this is sound practice cannot possibly be disputed; and the native farmer is already beginning to realize it.

The first problem was to find a suitable type of plough. In view of the fact that the native farmer has little or no capital, it was clear that his plough must be a very cheap one. The type of plough required by him is a double-breasted ridging plough. After a good deal of experimental work it was found that these could be made quite satisfactorily from wood, and furthermore that they could be easily copied by the native village craftsmen (see Frontispiece). There have in the past been several attempts made by administrative officers and missionaries to introduce ploughing into Northern Nigeria, and in each case the immediate cause of failure seems to have been that the type of plough was unsuitable; either because the European ploughs were not adapted to draught by bullocks, or else because they were single-breast ploughs. The wooden ploughs now in use can be made locally for less than £1. 10s. 0d., and they do quite satisfactory work. They have been used exclusively for all the ploughing on our experimental farms for the last three or four years, and those native farmers who have tried them have been well satisfied. As compared with the iron plough, they have one or two disadvantages. Their draught is much heavier. This does not matter in the light sandy soils around Kano; but it is a real disadvantage on the heavier soils of the Zaria district, especially in very wet weather. They are liable to shrink and crack in the dry weather, and care has to be taken to see that they are protected from white ants. Recently, a small English double-breasted iron

plough has been tried, and the results are fairly satisfactory. Its price is only about £3. 10s. 0d.; its work is
excellent; it cuts through weeds better than the wooden
plough, and even through the smaller roots of the cleared
bush, so that its use reduces the labour of stumping. But
it has been found that the shares wear down very rapidly
in tropical soils; and unless shares can be made cheaply
by local blacksmiths, the cost of their replacement may
make the use of the iron plough impossible for the
ordinary farmer. While experimental work on both
types still continues, the two are being demonstrated,
and the farmer can choose whichever he prefers. It may
be that the more wealthy farmer will like to have an
iron plough while the poorer man uses the wooden
one.

It is quite possible that the use of an iron plough may
in future make a cultivator unnecessary; but with a
wooden plough some form of primitive cultivator is
almost essential, at all events on any but very light land,
in order to break the land for ploughing in the early part
of the season and for weeding the growing crop. The
provision of a suitable cultivator was, however, never
a serious problem, as the native blacksmith can make
quite a satisfactory implement with a piece of bush
wood and a few bits of soft iron for tines.

The introduction of ploughs raised yet a further
question. The ordinary farmer, working with a hoe,
plants his crops on ridges which are between 3 feet
6 inches and 4 feet apart; but the type of plough which
two bullocks can draw can only make ridges from 2 feet
6 inches to 3 feet apart. The effect of the smaller ridge
on the yield of all the main crops had therefore to be

thoroughly investigated. This investigation took several years, but it was finally fully established that no serious loss of yield occurred.

The most important item in the system of mixed farming is manure, and this is a matter which has received the greatest amount of attention experimentally. The great objections to manuring by herding cattle on the land at night ("kraaling"), as already widely practised in Northern Nigeria, are first that the manure is applied in the dry season and several months elapse before it is dug in, and secondly the manure made by the cattle in the wet season is not utilized at all. Indeed the manuring season only lasts three or four months.

By keeping cattle in pens on the farms, wastage is avoided, manure can be made all through the year, and the bedding which is provided, when rotted down, adds to the quantity of humus-forming material which is available for application to the soil. For reasons which are not quite clear, the native farmer dislikes the idea of keeping his cattle under cover, and when he does so it is difficult to persuade him to supply enough bedding. Yet we have proved by trial that the cattle will thrive under a roof. For although on the experimental farms they have been so treated for several years, no difficulty has arisen. It might perhaps be argued that a covered pen would encourage such diseases as tuberculosis, but this can hardly happen if the cattle are turned out for several hours daily.

The value of the manure so made has been carefully compared with that made by kraaling, in actual field experiments; and over a series of years it was found that the manure made in 250 cattle nights by cattle kept in pens was equal in effect to that made in 400 nights by cattle kraaled on the land.

The Chief Veterinary Officer estimates that there are approximately 3,000,000 head of cattle including calves in Northern Nigeria. If these cattle were all kept in pens and supplied with bedding they could make approximately 7,000,000 tons of manure per year, which would all be available for application to the cultivated area. As it is, their excreta are equivalent in effect to 5,000,000 tons, but at least two-thirds of the manure is wasted in the bush. Thus it may be said that the equivalent of 2,000,000 tons at most goes on to the farms instead of 7,000,000 tons. To appreciate the significance of these figures it is necessary to realize that dressings of one to two tons of manure per acre per annum here are quite sufficient to maintain fertility at a level far above that of the present shifting cultivation. To the European who thinks in terms of dressings of from 10 to 20 tons per acre, a dressing of one or even two tons seems ridiculous; but under the influence of such dressings the experimental farms at Zaria and Kano have been steadily increasing in fertility. It is clear, from our repeated experiments, that under these arid sub-tropical conditions a dressing of only one ton causes a very remarkable increase in the yield of guinea corn; and that in the following year cotton benefits greatly from the residual effect of such a dressing. Why it should have so remarkable an effect is at present inexplicable; but there is no doubt whatever that it does. It is this very great result from a very little manure that makes mixed farming in Northern Nigeria mean bigger yields per acre as well as larger farms.

On the experimental farms we have used covered cattle pens, but it is possible that in our desire to conserve the manure to its fullest extent we have been unnecessarily careful; and that a walled but unroofed pen would be as good or even better. This point is now being studied experimentally, and if it is found that an open pen is satisfactory it may be easier to induce the farmer to take up the idea of supplying bedding.

Bullocks for draught purposes are readily available. The settled farmers who already own cattle usually have one or two male animals in their herd; and as these are all perfectly docile, whether castrated or not, there is no difficulty in training them for work. The big Fulani herdsmen dispose of their surplus males quite regularly at prices between £2 and £5. On a farm they appreciate in value even when worked; and the pioneer native mixed farmers have already made profits by selling beasts which they have had for a season or two, and buying untrained ones to replace them. It is not out of the question to work cows in the ploughs; but milk is such a valuable commodity that it would be unprofitable to do so if it causes a serious reduction in the yield of milk. This matter is the subject of experimental work now in progress.

The feeding of the working cattle has not proved a serious problem. During the wet season bush grazing alone will keep them in good condition, though it possibly pays to give them a little corn on the days when they actually work. For the dry-season feeding, the waste products from the farm, such as guinea-corn leaves and groundnut straw, would be almost enough if no cows were kept. We have always advocated the growing of a small area of *Mucuna* for hay and also the making of hay from bush grass as a supplement to the farm products. Hay can be made in October, which is a comparatively slack period, when haymaking does not make a serious drain on the labour supply. This is an entirely satisfactory time for making *Mucuna* into hay, but the bush grass has, by this time, deteriorated in feeding value, though the better species can still be made into hay which is at least edible. Silage made in the rains is probably more nutritious, and is being tried. But the feeding of the cattle will probably not be the problem on a native farm that the European is inclined to expect. On a native farm the working cattle do little or no work during the dry season, and even during the busy season are by no means overworked, while the cows give too little milk to need very heavy feeding.

Milk is a very valuable commodity among natives of Northern Nigeria, being much more highly valued than Europeans generally realize. With other prices at their present low level, milk is probably already the most profitable form of produce which can be produced on the farm: with better cows it certainly would be. It is readily saleable everywhere, and in the neighbourhood of Zaria its price is as much as 3*d*. per pint. The production of milk is therefore almost as important to the mixed farmer as the making of manure, and it cannot be too greatly stressed that ploughing is merely one among several advantages of mixed farming.

The ordinary native cow seldom gives more than 100 gallons a year, and as the Fulani are always reluctant to part with their best cows, it is difficult for the ordinary farmer to buy a really good beast, even if he is willing to pay a good price. As long as the native cows give such low yields little or no extra feeding is required. As in the case of the bullocks, grazing supplemented by the waste products of the farm, native salts, and perhaps a very little corn when they are milking at their best, is enough to keep such cows in good condition. In view, however, of the economic and dietetic value of milk, the Government of Nigeria has started a stock farm, the object of which is to attempt to improve the milking qualities of the local breeds by selection. When we are able to improve the native stock through the agency of pedigree bulls from this farm the real importance of feeding will first be felt. For a high-yielding cow will require something more than a mere maintenance ration.

It is surprising what a little extra feeding will do for even unselected animals. We are already beginning to find that the offspring of cows which have been well fed are considerably better than their mothers, and even the adult animals themselves usually increase their yield considerably after a year or two of reasonably good feeding throughout the year, instead of the "ups and downs" of the nomadic herd. For these reasons, the mixed farmer is being introduced to the idea of giving extra food to cows right from the start, and his small stack of *Mucuna* and bush hay, if not absolutely essential, is a step in the right direction. Simultaneously all practical foods and fodders are being investigated at the experimental farms, as well as the effects of mineral salts. It is hoped that results will be obtained which will in time be adopted by the mixed farmer as he gains experience.

It might be expected that the water supply would prove to be a difficulty during the dry season. This may indeed prove to be the case in certain areas; but such areas may well be left until the system has been fully developed in more favoured districts. In the meantime, the increase in the number of wells as a result of the use of modern methods of digging and lining will gradually increase the water supply of the more arid regions.

Apart from the epidemic diseases of rinderpest, pleuropneumonia and black leg, for all of which the Veterinary Department now undertake to give prophylactic inoculations, the only disease which has been at all troublesome is streptothricosis, a tick-borne skin disease which in an infected animal causes wasting and subsequently death. This disease can, however, be prevented by dipping with arsenical dips, and the same treatment will also cure the disease if it is taken in time. If, therefore, it is eventually found that this is a serious disease among the cattle belonging to mixed farmers, it may be necessary for the Government to establish dips at suitable centres. By charging a very small fee for dipping, these could probably be made to pay for themselves. It is, however, not unlikely that the farmer can cope with this trouble fairly well by his own methods.

In districts that are infested with tse-tse fly and the cattle disease that they carry, mixed farming of this type cannot be introduced. It seems just possible, however, that more use might be made of sheep or goats in direct connection with farming. At present, though they browse over the fields after harvest, no fodder is grown for them and their manure, poorly conserved, only suffices for the small special crops near the house. A system of mixed farming based on the growing of beans, feeding them to sheep and using the manure more effectively, might possibly prove profitable where cattle cannot be kept. In spite of some obvious arguments that can be adduced against it, the proposal has been deemed worth a trial, and experiments in this direction are now being undertaken.

CHAPTER 8

EXTENSION WORK

THE best principles to be followed in the extension of improved agricultural methods is a fruitful subject of discussion in most countries, and especially in those of which the people are illiterate or but slightly educated.

The extension of agricultural practices involves two measures: first the introduction of the new crop, variety, or practice to the farmers, and secondly the steps that are necessary to make its adoption possible or profitable. These latter measures are often very important and may involve the organization of a supply of the new seed in a pure condition; or the control of marketing, so that the pioneers may not fail to obtain the value of their new type of produce because the quantity is at first too small to attract the attention of the trade; or the provision, usually through the agency of co-operative societies, of the capital that the farmer needs for the new undertaking. These measures are often matters of some difficulty, and they are unquestionably an essential part of the functions of Government in such countries as West Africa. Such measures may perhaps seem to some to be outside the functions of Government, and they certainly are liable to lead to an undesirable diversion of the energies of the Agricultural Department into the business of seed supply or the organization of markets. But they are generally inevitable. It is, for instance, quite useless to ask a farmer to grow a new superior variety of export crop merely because it will command a higher price than the ordinary, if, in course of time, it should completely replace the ordinary. Such a distant conditional promise is of no use to anyone; it is necessary to assure the farmer that measures will be taken to obtain the proper price for him next harvest, however small may be the total quantity of the new superior type. The lack of such measures has in the past led to the failure of many "improvements". But the measures needed in each instance, and in each country, are peculiar to the special circumstances, and can hardly be usefully discussed in general terms. It is proposed to deal here chiefly with the steps to be taken in introducing the new variety or method.

The oldest method might be described as that of universal official propaganda. The essence of this method was that the European officer decided what it ought to be profitable to the native farmer to do, and then officially advised all farmers to do it. Thus it might be decided that it would be profitable to the people of a certain district to grow American cotton, or rubber, or sisal, or cocoa or whatever it might be. It did not matter whether the farmers thought it would be profitable to do so: they could not be expected to look so far ahead; the European must decide for them. American cotton growing, or rubber planting, became the official policy to be encouraged by every possible means. Seed was obtained and given away free and the whole influence of the administration was used to encourage the farmers to plant it—sometimes successfully; but sometimes the farmers proved remarkably reluctant to follow the good advice so generously offered.

Unfortunately, this method has the merits of a glorious simplicity, and of appealing directly to the best instincts of the well-meaning officer. He is doing his duty as the enlightened leader of a backward people. If he succeeds it shows how he can "get to the back of the black man's mind", and it demonstrates his personal influence; if he fails, it proves what difficulties he has to encounter in the innate conservatism of an uneducated people; and anyway he has done his best, and it obviously could not be suggested that he has done any harm to anyone. The farmers do not pay rent or wages, so the unsuccessful efforts of the trusting pioneers have not cost them anything. It does not occur to the officer that those who take his advice may not believe in it at all, but may merely be obedient; and that any confidence they might have in the white man's wisdom is very easily shaken. Except in a very few cases, such efforts have always failed as soon as the propaganda has slackened. West Africa is littered with the failure of past efforts of this kind, while the farmer still goes quietly on farming his own crops in his own way. In the few cases where success has been achieved by such means, it has been because the new practice or crop did accidentally happen to suit local conditions, and was a practical paying proposition. It was a hit-or-miss system, and there were many more misses than hits.

The new or modern method has been simply stated in the departmental instructions of the Nigerian Agricultural Department as follows:

"(a) No recommendation should be made to any native directly, nor to any political officer, or other person who may convey it to any native, which is not based on the results of experiments.

"(*b*) When such a recommendation is made it should never be made in such a way that it may take the form of a general instruction; but it should be made (of course with the concurrence of political officers) to certain selected natives or in certain selected villages. If it proves successful there, the example will certainly be followed by others. If the soundness of the advice is thus proved, general recommendation might thereafter be justified, but will almost certainly be found to be unnecessary, or at all events avoidable".

The first of these two instructions has already been discussed in Chapter 1 and does not concern us here, although it has entirely altered the local interpretation of the term "extension work". For the latter now no longer means addressing meetings and exhorting the farmers to do what the Europeans think it will pay them to do. Instead the attitude adopted is that we have proved that so-and-so should be sound for the native farmer so far as we can judge from our experiments; the next step therefore is to induce a few people to give it a trial themselves, realizing that it is only a trial, and that we welcome their criticism.

There seems to be a marked difference between the native farmers of West Africa and those of more advanced countries in their response to "mass methods". In advanced countries, extension lectures by experts, the visits of large parties to experimental farms on "farmers' days", special public demonstrations, farmers' debates opened by an expert, and similar means apparently lead to the desired ends of stimulating interest and spreading knowledge. But, as a matter of practical experience—and such methods have altogether been fairly widely tested—the same is not true with peoples such as those of West Africa. One difference undoubtedly arises from the motive which brings the audience together. In advanced countries the only motive that will bring them is genuine interest, in other words self-interest; the hope of learning something that can be promptly turned to a profitable account. In West Africa the audience may come merely because they are in effect ordered to do so, or out of politeness or curiosity. It would be easy to explain the failure of lectures in West Africa as due to the fact that the lecturer speaks in a language that is not his own, or through an interpreter. But it would seem that there is some more fundamental reason for the failure than this. It appears that to take advantage of a lecture or demonstration—never mind how simple it may be—needs a greater degree of general education in the members of the audience than a European is inclined to suspect. One who has been used from childhood to learning in class rooms and lecture halls, cannot appreciate the degree of mental training which is required before one can learn in this way. It is not so much a question of degree of intelligence or of interest, but of training in using intelligence in a particular way. If a party of African farmers is taken round a farm, it does not seem to matter how much explanation is given, or how much the important points are emphasized, the points that the visitors will especially observe and remember are the regular shape of the plots and the straightness of the lines of plants. If demonstrations of grafting are given, the shape of the knife used will attract most interest. Invariably some entirely unimportant point will seem to monopolize interest, and apparently even be regarded as the essential point.

The use of the cinematograph in agricultural extension work among primitive people has been frequently discussed though but little tested in practice. At first sight such a method might appear suitable for illiterate people, but it is probable that it would be found in practice to be subject to objections similar to those mentioned in the last paragraph.

In most countries with a peasant population, trials have been made of small demonstration farms. By this is meant a Government farm of an ordinary native size which is not used for any experiments, but is merely conducted by the best approved methods, with the object of convincing peasants in the neighbourhood that those methods are practical and profitable. The results from such farms have been very variable; sometimes they seem to prove effective agents in extension work, sometimes they seem to be quite useless for this purpose. One great difficulty is to convince the people that the farm is profitable. If the crops are not very good, the farm fails in its purpose; if they are very good, the neighbouring peasants naturally think; "that is all very well, but you are spending more on labour and manure or appliances than we could afford. They are good crops, but they can't pay". The demonstration farm may be paying well, returning what is spent and paying for depreciation on the superior implements and buildings and stock; though it must be admitted that that is not always the case. But even if the farm does pay well, how is a population that cannot understand a balance sheet to be convinced of that? All that is possible—and this is unquestionably a point of fundamental importance in connection with such farms—is to avoid the erection of better buildings than are absolutely essential, or the use of one single unnecessary tool or appliance. Everything on the farm must be ordinary, and not only such that it would pay the farmer to use, but also the cheapest possible, such as he would be likely to use. All non-essentials, all "frills", must be most rigorously eschewed. Otherwise the farm will inevitably fail in its purpose, no matter how good the crops, nor how satisfactory its balance sheet may be.

Another essential is that the farm must have something to teach. This sounds simple and obvious, but there is more in it than might be thought. In an English countryside there are always one or two men who are universally recognized to be exceptionally good farmers from whom others can learn. This is hardly the case among "native" farmers. Their methods are far more traditional; the individual farmer's knowledge is the knowledge of the tribe and everyone has the same knowledge, neither more nor less. So far as one man is a better farmer than another it is mainly because he (or his wife) works harder, not because of greater knowledge, skill or judgment. Similarly, in a more advanced community, the demonstration farm may successfully demonstrate general good farming, to show that such good management pays. This will not suffice in West Africa. The demonstration farm, if it is to succeed here, must have some obvious superiority in method over its neighbours.

Finally, success probably depends, more than anything else, on the personality of the manager of the demonstration farm. If he is not too superior in social standing to his neighbours, yet is a man of sufficient intelligence and character to command their respect, he may be able to make intimate personal friends among them and to acquire real influence. If this is not possible, and he has instead to resort to inviting parties formally to inspect his farm, then little success will be achieved.

On the other hand, the extension worker in West Africa has some advantages that his counterpart in more advanced countries has not. In the first place, there seem to be always in any primitive community, no matter how primitive, some individuals who are remarkably ready to give a trial to a new method if they are approached in the right way—or indeed if they are not approached at all; for nearly every extension worker who has had any success can recount instances of men who have come in, often several days' march, because they have heard about some new method that was proving successful, and wished to look into it with a view to giving it a trial themselves.

There is also another characteristic of the West African peasant that can be most useful in this connection. In England the prophet of agricultural progress is liable to find himself without honour in his own village; no one would have more difficulty in convincing his neighbours than a young man of the village who has been to a course at an Agricultural College. If he becomes a leader in that village it will be only by virtue of practical success, and in spite of, rather than because of, his belonging to the locality. This seems not to be the case in more primitive communities; if one man can be convinced, others will readily follow him. This fact can be made use of in two ways. Native assistants employed by the Agricultural Department, if they are really themselves convinced of the soundness of a new method, can generally find some of their relations and tribal connections who will give it a trial on their own farms; and every effort should be made to use the assistants' influence in this way, in their personal unofficial capacity. Again, thanks to the fact that others of the same tribe will follow one who succeeds, the worker can concentrate on helping effectively the few specially enterprising men whom he finds or, still better, who find him. A successful farmer is the best demonstrator to farmers. The official workers, both native, and still more European, may be so few in number that they can only reach directly a minute fraction of the population; but if those whom they do so reach succeed and are truly convinced, then others will soon be reached indirectly.

It is well that it is so, for it is only in the introduction of the simplest and easiest improvement, such as the adoption of a superior variety of a crop which is already grown, that the pioneer can be merely advised to give it a trial and then left to his own devices. When the innovation is of a more complicated nature he must be regularly visited and helped with advice, criticism or encouragement. This is not only because he may make mistakes. It must be remembered that among a people where everything is normally governed by tradition, a pioneer needs moral support. For he is going against public opinion and facing the possibility at least of ridicule; and since he is associating himself with the officials of a Government that collects direct taxes from farmers, his action is even liable to be regarded with hostile suspicion. At the same time, it is a mistake to harass him with too frequent visits, too much criticism and too many suggestions, to "spoon-feed" him too much. It is for these reasons that success in extension work depends so greatly on continuity in the official personnel, and on the personality, judgment and discretion of the officials, both European and native. Nothing permanent will be achieved by an officer who is in one district for two or three years and is then replaced by a different individual; nor by one who thinks that he only has to give advice and the farmer will of course be delighted to follow it.

A problem peculiar to Africa arises in connection with the relationship of native chiefs, the Emirs and District Heads, to the agricultural extension work. These are the most influential men of the district, and it is natural that the administrative officers, if not also those of the Department of Agriculture, should turn first to them when attempting to introduce any change. They have influence, and it is natural that the agricultural officer should wish to utilize it. The administrative officer further wishes to recognize and enhance the chief's influence, by such prestige as the latter may attain by acting as a leader. Indeed it often seems that the administrative department thinks it entirely wrong that the agricultural officer should approach individual farmers directly and try to make friends with them personally, instead of working wholly through the native chief; for to do this savours of neglecting the chief's position and of undermining his authority. The administrative officer would not act thus himself and he does not see why the agriculturalist should wish to do so. Moreover, so great is the authority of some chiefs, so persistent the relics of the feudal system in some places, that the people will not presume to take up a new practice until the chief has clearly shown his approval of it by adopting it himself.

On the other hand, there are many objections to using the chiefs as the first pioneers of new agricultural practices. There is doubtless much that is democratic about the chiefs' attitude, but still they are essentially autocratic; and to many of them, if it is desired that some practice should be adopted by the people, the natural thing is to give an order that it shall be taken up, say by one farmer in each village. This is fatal to any hope of success; and one such occurrence can handicap future progress for years. Again some chiefs are personally popular, but not all; and to work hand-in-glove with an unpopular chief is hardly the way in which the agriculturalist is likely to make real friends among ordinary individual farmers.

Further, in the early stages of the introduction of the new practice there is still a large element of the experimental—it is necessary to see whether the native farmer really can adopt it, or whether it involves some unforeseen difficulty for him. To see how a chief or district head gets on is of little use. He may, owing to his greater means or superior intelligence, succeed where an ordinary farmer would fail; or he may fail where an ordinary man would succeed, because the chief is generally, in effect, either an absentee landlord, or a gentleman-farmer who leaves his farm entirely to a bailiff. For similar reasons, such a man's personal opinion is often hardly representative of the farmer's opinion; even apart from the possibility of his giving opinions that are biassed in whichever direction he judges will be the more acceptable to his superior administrative officer.

Yet it is not feasible to leave the chiefs entirely on one side; and various compromises have to be effected. The agriculturalist must continually repeat that he does not want people pressed, or "persuaded" or "advised" to try his recommendations. It is easy to obtain a verbal acceptance of this idea, but almost impossible to get it followed in practice, and constant repetition is necessary. Again it can sometimes be arranged that a near relation of the chief takes up the new practice first, instead of the chief himself; at the very least, every effort should be made to ensure that the chief acts in his private capacity as a farmer, and not in his official one. Above all, the agriculturalist must remember that his object is not to get chiefs to adopt his practice, and their conversion does not in itself constitute progress; that is only a necessary sideline; the agriculturalist's real object is to get into direct touch with some ordinary farmers as individuals and as unofficially as possible; progress is to be measured only by the ordinary farmers who adopt the new method for no other possible motive than that of enhanced profit.

One of the great difficulties in West Africa is the provision of capital for any improvements in farming; for many improvements, and those often the greatest, can hardly be taken up by men without any capital. Before he can start mixed farming or use small machines for extracting palm oil and kernels, a man must have the money to buy the cattle or the machines. Even to plant cocoa or oil palms and bring them into bearing, capital is needed if one man wishes to handle a relatively big area, say an acre or two acres all at once.

The capital required is very small; £5 or so is all that the ordinary progressive peasant would generally need. Improvements needing more capital than that cannot at present be contemplated for the ordinary farmer. But even £5 is more than he generally has. The fact that the farmers of West Africa have generally no clearly established and recorded individual title to their land or to its use, rules out loans on mortgage. This is undoubtedly a restriction upon the bigger men, but it is arguable whether the ordinary very small ones would be better off in this respect if they had clearer titles. Commercial banks cannot deal with such petty mortgages, and experiments in various countries have shown that even an official "Land Bank" cannot successfully deal directly with individual peasants. If such small peasants could obtain mortgage loans privately it would be only from small native capitalists, on exorbitant terms.

The apparently obvious solution of this difficulty, as indeed of many others connected with agriculture among peasants, would seem to be found in co-operative societies. There is very good reason to believe that the fundamental basis for co-operation—the personal honesty and integrity of the peasant, his willingness to pay what he owes—will be found to be present to at least as high a degree in West Africa as anywhere else. Moreover, the democratic nature of the old tribal organization in many parts, the habits of forming guilds, age classes or secret societies, and of village government by council of elders—all these help to provide a substratum of what may be called "co-operative mentality". That the conditions are exceptionally favourable in this respect seems to be proven by the success in the Gold Coast and in Nigeria of Cocoa Fermenting and Sale Societies. In India and Ceylon co-operative credit has proved relatively easy, and in India progress in this direction has been very rapid; but yet, after twenty years of co-operative work, there have been in either country very few successful peasant societies for the sale of produce; and it seems that generally the peasants of those countries have not yet developed co-operative character and ability enough to undertake these more difficult operations. That West Africa should start with a difficult type of society and meet immediately with a considerable degree of success would seem to indicate a special natural aptitude in the people.

Experience in India and Ceylon has shown that illiteracy and lack of education are no obstacles to co-operation, at all events for credit, so that there should be no difficulty on this score in West Africa.

There is, however, one great difficulty which is more or less peculiar to West Africa. Experience in other countries has shown that to connect co-operation with the local administration, as for instance by making the local village or district head the president of the co-operative society, spells certain failure. Yet, except in the most advanced parts of West Africa, it is hardly possible for co-operative societies to be entirely unconnected with the native administration, and not seem to be an influence that is subversive of that of the administration. It cannot be doubted that the native officials would resent such independent organizations, and oppose them; and, in that event, they could almost certainly prevent their succeeding. It is only in the most advanced districts, if indeed there, that the native authorities are sufficiently broadminded to be able to adopt an attitude of benevolent neutrality, or to provide support without interference to co-operative societies not under their control. In Northern Nigeria, for instance, such an idea can hardly be contemplated at present. Yet it would seem that this is a difficulty which must eventually be overcome; for the West African colonies, like other peasant countries, will probably eventually find that there is a limit beyond which further material progress is only possible through the agency of co-operative societies. In this respect, as in many others, the future of the country largely depends on the ability of the native administrations to change gradually into more modern and liberal institutions, without losing their authority. Fortunately the more enlightened native officials are of a far less conservative turn of mind than seems commonly to be believed, and there is thus reason to hope that the native administrations of the future will eventually become so strong and enlightened as to be able to welcome any organizations, such as co-operative societies, which can strengthen the peasants' position.

PART II
CROPS & STOCK

CHAPTER 9

THE OIL PALM (*ELAEIS GUINEENSIS*)

THE oil palm is indigenous to West Africa, and, until quite recently, was grown nowhere else. It occurs in all the countries from Sierra Leone to the Congo, wherever the average annual rainfall amounts to 60 inches or so distributed over not less than eight months of the year, and where there are many people. For, although it grows semi-wild, the oil palm is always associated with population. There are indeed some areas which are very thinly populated at present where the oil palms are numerous; but it is probable that those areas were more densely peopled in the past.

Palm kernels form the bulk of the exports of Sierra Leone and constitute the most important among the several exports from Southern Nigeria, while palm oil ranks second. There is a very considerable export of both oil and kernels from the Belgian Congo, and lesser amounts from other West African countries.

Recently the cultivation of the oil palm has been taken up by the planters of Sumatra and, in a lesser degree, in Malaya. As yet the export from these Eastern countries is quite paltry in comparison with that from West Africa. But a few years ago it was feared that the extension of oil plantations in the East might be so rapid as to exceed the growth of the world's demand for palm oil, and thus to cause a serious diminution of its value in the world's market. For although the world's total demand for vegetable oil is very considerable, and although to some extent one oil can replace another in manufacturing processes, yet there is also a special market for each particular class of oil, and a great over-production of palm oil would be likely to lead to a heavy slump in its value. However, it seems that the planters in Sumatra and Malaya have recently realized that their early estimates of the yields which they could expect to obtain from cultivated oil palms were much exaggerated. Thus, although the cultivation of the palm in the Far East will almost certainly continue to spread steadily, there now seems less danger than there was a few years ago of a very rapid extension of these plantations. Yet it still behoves the West African native producer to improve his methods, not only in order to be able to maintain his production in the face of the eventual competition of the Eastern plantations, but also further to increase his own wealth, especially as the recent low price of rubber is again stimulating oil palm planting in the East.

The importance of the oil palm in Southern Nigeria cannot be over-estimated, and the position is very similar in Sierra Leone. In the palm oil districts of Southern Nigeria, palm oil and palm kernels are the only commodities that the people produce for export, and they are their sole source of income. Since they grow their own food, they might not starve if they could not export palm oil and palm kernels; but, if life means anything beyond avoiding starvation, their life depends entirely upon the oil palm. The soil of the great central portion of Southern Nigeria from Benin nearly to Calabar, which constitutes the great palm oil region, is peculiar and poor; as yet no alternative export crop has been found which could replace the oil palm on that soil. The costs of caskage, transport and handling of palm oil are so high, that if the price falls very low, there would be nothing left at all for the producer of the oil. In the latter part of 1930 and the early part of 1931, it appeared as though this stage might be coming in sight.

On the other hand, a consideration of all the factors involved seems to point to the conclusion that if he could improve his methods to the full extent of which he is capable, the Southern Nigerian native could always produce palm oil as cheaply as anyone else in the world. In other words, any price at which plantations could produce at a profit would be profitable to the Nigerian native. Palm oil and kernel production may be liable to be endangered by the fluctuations of the world's supply and demand; but there are few, if any, products which are subject to less violent fluctuations of price. The difficulties in the way of finding and developing any big new industry in this area are very great. Thus the best way to safeguard the future of the palm oil belt of Southern Nigeria seems to be to try to improve that industry, rather than to try to replace it by a new export crop. The native of Southern Nigeria suffers disadvantages from lack of capital and the ability to use large-scale methods; so that he cannot hope to rival the Eastern plantations in efficiency in all respects. Yet he could greatly improve on his present methods. The peasant has the advantage of the incentive that comes from independence; and this, if he improved his methods as far as he is able to do so, might well outweigh his disadvantages.

The palm oil industry in Southern Nigeria is susceptible of improvement in respect of the way in which the trees are grown, the variety of the palm tree that is grown, the proportion of the oil that is extracted out of the fruit as compared to the amount which the fruit actually contains, the quality of the oil prepared, and, in a lesser degree, the method of extracting the kernels from the nuts. The discussion of each of these questions in turn will include reference to most of the important facts that have been discovered about this tree and the preparation of its products, in the course of the ten years' close study of it that has been made in Southern Nigeria.

At present, in Nigeria, the oil palms occur mainly in one of two ways: scattered on farm lands, or in groves. The palms in the farm land are usually individually owned, but those in the ordinary groves are generally communal property. There are also usually many palms growing between the houses in the villages, sometimes on such a scale as to constitute a "village grove". These palms are always individually owned, and generally attain distinctly better development than those in the farm lands or ordinary groves. If a young oil palm grows up among a tangle of "bush", it makes slow progress, forms but a thin trunk and a small head of leaves, and will not begin to bear fruit until it is some 30 feet in height. Should the bush be cleared from around a young palm (less than 30 feet in height) and be kept clear, the tree will respond to the treatment and begin to yield fruit; but if the palm has formed any trunk at all it will never attain anything approaching the same size or yield as it will if it is originally planted in cleared land and kept free from excessive weeds throughout its life. From these considerations it would appear that the palms which grow on the farm lands should be superior to those in the groves, since they have not to compete with dense "bush". Sometimes it is so; but under the system of shifting cultivation the land repeatedly reverts to "bush" for long periods, when the palms in it naturally suffer; and when the bush is cleared and burnt the palms are liable to be damaged by the fire. Moreover, an oil palm growing in a farm—especially a palm of a few feet in height—has an obvious detrimental effect upon the crop in its vicinity. The farmer therefore frequently cuts most of the leaves off the palms of this size, which again greatly checks their development.

Palms which are planted in cleared land at an adequate spacing (say 28 or 30 feet apart) not only make much bigger and heavier yielding trees, but present also an early and very important advantage in that they begin to bear fruit at four or five years old, will be bearing heavily at seven years, and give a maximum yield from ten years onwards. At four years the palm has hardly formed a trunk at all, and for several years the fruit is still within reach from the ground, and again for many years after that the fruit can be harvested with a ladder. The wild palm, on the other hand, usually only begins to fruit when it has reached the height at which climbing becomes necessary. When the tree has to be climbed, harvesting is slow and heavy work, so that the early fruiting of the planted palm is of great practical importance.

From the results of investigations, it is clear that if the palms on every acre of ordinary palm groves and in farm lands in Southern Nigeria could be replaced by an equivalent area of planted palms, the yield of fruit would be increased threefold, or even more, and the labour of harvesting would be vastly reduced.

When the Nigerian Agricultural Department was reorganized in 1921–2, it was immediately realized that the introduction to the native farmer of the idea of making small plantations of oil palms was one of the most important pieces of work that could be undertaken. But great difficulty was experienced in obtaining land on which to conduct experiments or demonstrations in the palm oil belt, or in finding any families or small communities who would make such experiments themselves, or allow them to be made on their land. It was only in 1927 that this difficulty began to be overcome in two or three districts.

There are two distinct methods by which a native farmer can make an oil palm plantation. The obvious method is simply to clear land on which there are no palms and to plant them. But where there are many palms of various ages growing among "bush" as in the ordinary communal palm groves of the south-east of Nigeria, it is more profitable to retain some of the old palms to give some return while the new young palms are being established. To distinguish this process from the simple planting of a plot of palms, the term "Palm Grove Improvement" is used.

Experiments conducted by the Agricultural Department show that a young oil palm is remarkably hardy as compared to most "plantation" trees. It is thus quite unnecessary, and indeed undesirable, to clear and stump the land thoroughly before planting palms. It is only necessary to clear the growth above ground, and to prepare planting holes. Thereafter the young plants must be ring weeded several times a year and the regrowth of bush cut back some three times a year, until gradually it is mastered by the growing palms. Alternatively food crops can advantageously be grown between the young palms; and experiments show that the palms do not suffer at all thereby. Naturally, as the palms grow, the amount of food that can be produced between them becomes less and less, and it is not worth while planting anything after three years. In these ways palm plantations can be established by the Nigerian farmer remarkably cheaply. At the end of 1931 there are some 340 acres of such plantations, belonging to 99 individuals. The movement, small as it is at present, is spreading with increasing rapidity.

Where there are ordinary groves, consisting of mixed palms and bush, the plan of palm grove improvement has been to cut down the bush, to thin out the old palms, naturally selecting the best for retention, and to plant seedlings in the spaces between them, or to leave a natural seedling when there happens to be one that is young enough just in the right place. Ten such experiments were begun in 1928, each experiment consisting of several cleared half-acre or acre plots and several alternating uncleared plots. On the existing trees, the effect of the clearing depends upon the number of young trees left that are less than 30 feet in height, for palms do not respond appreciably to treatment after this height. The effect of the thinning and clearing has been to increase the yield *per acre* by amounts which average 100 per cent., but vary from 20 to 200 per cent. in the different experiments; the treatment has nowhere failed to increase the yield per acre in some degree, in spite of the cutting out of many old palms. While the old palms in the cleared plots have been thus giving an increased yield, the planted palms have grown satisfactorily—not quite so well as in a plantation pure and simple, but still satisfactorily. Now, after four years, the critical stage is approaching when it will be necessary to cut down the old palms in order to allow the young ones to develop fully; and the cutting out may possibly result in a reduction of the yield for a year or two. It remains to be seen whether in view of the family ownership of these experimental groves, the owners will be able to bring themselves to take the drastic step of cutting down good old palms.

Naturally, as the young palms are only just beginning to bear, the full benefits of the measure are not yet visible. Actually, the first result of these demonstrations has been to stimulate, not "palm grove improvement", but the making of small plantation plots. The communal ownership of the groves is a great handicap to their improvement. The experiments have already indicated, however, that the groves can be replaced by planted palms in the manner described, and since the final result must be such a vast improvement on the old groves, it would seem that eventually some means must be found of overcoming the obstacles in the way of the general adoption of this important measure of permanent improvement.

Owing to the fact that the oil palm is cross-fertilized and that each generation takes several years before fruiting, knowledge of the laws of inheritance of the desirable properties in the fruit is still lacking; indeed it is still questionable which is the best type, and how far the superiority of the trees in the better "palm belts", e.g. the south-east of Nigeria, is due to hereditary characters, and how far it is due to the soil or climate. It must suffice to say that it is evident that superior quality of fruit and heavy yield is at least partly hereditary, and that it is clear that a great increase in yield could be effected by breeding and planting superior types. Self-fertilizing and crossing of good trees is in progress, and seedlings produced in this way are available for native farmers who wish to make plantation plots. It will be many years before the "artificial" seed will be pure and produce trees of a uniform standard of excellence; but there is reason to hope that on the average those produced even now will be superior to trees from ordinary seed.

The native processes for the extraction of oil are very variable, but can be roughly divided into "hard-oil" and "soft-oil" methods. In the former the fruit is softened by being fermented in heaps and mashed, and the oil partly runs out and is partly squeezed out of the softened mass. In the soft-oil process the fruit is softened by boiling and pounding, then the oil is separated by working the mass in water until the oil rises and can be skimmed off. The hard-oil process yields from 55 to 65 per cent. of the oil in the fruit according to the quality of the fruit. The soft-oil process yields less, say 55 per cent. of the oil in fruit that would give up to 65 per cent. by the hard-oil process, but it is of better quality. The soft-oil process is employed chiefly in the Western Provinces of Nigeria where much of the produce is intended for internal trade; the hard-oil process is used in most parts of the Eastern Provinces where the bulk of the oil is for export. The soft-oil process takes longer, but can largely be carried out by women; the labour of the hard-oil process, though it occupies less hours, is generally done by men.

Now by the use of heavy machinery it is possible to extract some 85–90 per cent. of the oil in fruit that would yield only 65 per cent. of its oil if treated by the hard-oil process; and by the use of solvents it would be possible to recover nearly 100 per cent. of the oil, though it is not profitable to do so at present prices. Such efficiency can only be attained if very large quantities of fruit are dealt with in large factories under highly skilled technical management. Thus, in this matter, the Nigerian extractor is necessarily at a disadvantage as compared to the Eastern plantations. Yet even here the small extractor can improve his methods. Several years' work by the Agricultural Department in Nigeria has resulted in the evolution of a process based on the use of a small press which can be worked by hand. By this means, it is possible to extract fully 80 per cent. of the oil in fruit from which at most 65 per cent. could be extracted by the hard-oil process, and the oil is superior to that of the soft-oil process. As regards labour, the press process is at least on a par with the native methods. Over twenty of these presses have been bought by native extractors, and it is encouraging that no one has ever adopted the process and subsequently given it up. There is no doubt that if a good press could be produced at £5 or even £8, it would soon be adopted in large numbers. So far, however, the cheapest type offered to the native has cost £17. 10s. 0d., though a cheaper one is about to be produced. It is only the price and the lack of capital that hinders the adoption of these presses in many districts.

Quality in palm oil, as apart from purity, depends upon the relative "softness" of the oil, which again depends on freedom from free fatty acid. Soft oil contains relatively little free acid, hard oil may consist of as much as 60 per cent. of free acid. At times "Lagos soft oil", with 12-18 per cent. of free fatty acid, has been worth as much as £5 per ton, or 20 per cent. on the local value, over the price of hard oil. Soft oil has been more valuable than hard in the past chiefly for two reasons. One reason was that when soft oil is used in soap manufacture it yields more glycerine as a by-product than does hard oil. The other reason is that efforts have repeatedly been made during the last ten years to use palm oil in the manufacture of margarine, for which purpose only soft oil is suitable. There is, however, an objection to the use of palm oil for this purpose; for, although the peculiar taste of palm oil can be removed in a margarine factory, it is found to return again if the consumption of the margarine is delayed for a few weeks. At the moment it happens that glycerine is relatively very cheap, and palm oil is apparently not being used at all in margarine making. Soft oil, especially the oil with only 6 or 8 per cent. of free acid, still commands a small premium, because it alone can be used in the tin-plate industry, which constitutes one of the minor markets for palm oil.

Oil which contains only 4 or 5 per cent. of free fatty acid can be prepared equally well either in the large machinery of a big estate, or by the Nigerian small extractor who uses a press. The oil, prepared by the ordinary native soft-oil process, containing usually some 12 per cent. of free fatty acid, will command a premium because of its higher content of glycerine when that substance is dear, but is not suitable for the margarine factory. Oil containing only 1 or 2 per cent. of acid can be prepared by purely native methods, and sometimes is so prepared in small quantities when the extractor proposes to eat the oil herself. These very low percentages can only be attained by any process if none of the fruit is allowed to become overripe.

The use of small hand-power or treadle machines for breaking palm nuts to extract the kernels is slowly spreading in Nigeria, and there are perhaps some fifty such machines in use. Neither the quantity nor quality of the kernels extracted is affected, but the time taken is greatly reduced, though the work is of course much harder.

Palm kernels are pressed in England and on the Continent; they contain 50 per cent. of their weight of oil. The oil is roughly similar to coconut oil or groundnut oil, and is used in the preparation of margarine or of culinary oil. The remaining cake is used for cattle food, though in England it is seldom used alone, but is incorporated in proprietary brands of cattle food.

All palm oil and kernels exported from Nigeria and kernels exported from Sierra Leone are subject to Government inspection. The standards, which are very rigidly enforced in Nigeria, are that the impurities shall constitute less than 2 per cent. by weight in oil and 4 per cent. in kernels.

CHAPTER 10

COCOA, KOLA, COCONUTS, RUBBER

COCOA (*THEOBROMA CACAO*)

IN British West Africa cocoa is grown only in a part of the Gold Coast and in the south-western part of Nigeria. The tree needs a climate which is humid for the greater part of the year. If the dry season is long or very severe cocoa will not succeed: its sensitiveness in this respect is clearly seen in places on the borders of the cocoa belt. There cocoa will flourish in any small valley where the trees are sheltered and the atmospheric humidity conserved a little in the dry weather, while, if planted on level ground nearby, it soon dies. Again, investigations in Nigeria indicate that cocoa is very sensitive to soil acidity, and that this factor practically rules out cocoa cultivation in a large part of Southern Nigeria. On the soils of the "Delta" provinces, which are commonly highly acid, many unsuccessful attempts to grow cocoa have been made in the past by native farmers and on the Government farms. Here and there, where there is a patch of soil that is less acid than usual, the trees may live for a few years, and bear a small crop; but more commonly in this area they die even before coming into bearing. It is just possible, though not very probable, that cocoa might be grown successfully on this soil, if it were periodically heavily limed, and this is now being tried experimentally.

The native cocoa planter does not clear and stump the land on which he proposes to grow cocoa so com-

pletely as is customary on estates. But even he stumps
the land more thoroughly for cocoa than he does for
other crops; and experiments on our Government farms
in Nigeria confirm that this is necessary. When possible,
forest land is used for establishing cocoa plantations,
because there is a less vigorous regrowth from forest
stumps than from those of secondary bush. After the
land has been cleared, food crops are planted, and cocoa
seeds sown "at stake" through the food crops. Suc-
cessive food crops are grown for three or four years,
and are weeded when necessary; but no cultivation is
given specially for the young cocoa. By the time the
cocoa trees have formed a complete canopy over the
field, that is after four years or so, the stumps of the
original bush will have been killed by the repeated
cutting off of the small shoots which they put out. If
the food crops compete with the cocoa to some extent,
the cultivation which is given to them keeps down weeds
effectively; and their presence protects the soil from
wash. It is doubtful whether the cocoa would be any
better, or even so good, if the food crops were not grown;
and, naturally, if the latter were not grown, it would
cost very much more to establish the cocoa.

The cocoa trees on a native farm in West Africa are
almost invariably much too thickly planted; they are
commonly not more than 6 or 7 feet apart. Such close
planting is possibly advantageous in the first year or two
after the young trees come into bearing. But experiments recently carried out in Nigeria show that the yield
of such fields, when only a little older, can be increased
by cutting out many of the trees. It would seem to be
better, in the long run, to plant the trees wider apart at

first, and the native of Nigeria is beginning to come round to this opinion.

Cocoa trees on West African native farms, once established, ordinarily receive little or no further attention; and in view of this fact, and of the close planting, it is very remarkable that the trees in Nigeria are practically free from diseases due to either insects or fungi. For in most countries cocoa is a somewhat delicate tree much subject to pests and disease. In the Gold Coast there is a considerable amount of root disease, and much damage is done by insects, especially one called Sahlbergella. In both countries, the crop is considerably diminished by the " Black Pod Disease ". This especially damages ripe pods, and it is believed that if every pod could be picked as soon as it is ripe, the disease would cause little loss. It is not practicable for a farmer to pick each pod immediately it is ripe; but it would be feasible to harvest rather more frequently than native farmers in West Africa commonly do; and even this would greatly lessen the losses from this disease.

It has frequently been suggested that the disease could be kept in control by the early removal of all diseased pods, and of the many redundant young fruits, which die while they are still very small: the burning or burying of all shells of the pod (husks) after harvesting has also frequently been recommended. Scientific investigations into these points have been made in the Gold Coast; and in Nigeria field experiments have been carried out in which the yield from plots most thoroughly treated in this way has been compared with the yield of untreated plots. The results in both countries go to show that these measures are quite valueless in practice.

The Nigerian experiments, however, seem to show so far that the proportion of diseased pods is somewhat reduced if a thick plantation is pruned of dead and excessive wood, and the redundant trees cut out. This process must not, however, be carried out too vigorously or too completely in one operation, for if the shade canopy of the trees is greatly reduced, the attacks of insects would be encouraged. These experiments are as yet hardly numerous enough to be conclusive on this point, and they are being continued and extended. Experiments are also being carried out to test the value of various fungicidal sprays and dusts; but it is unlikely that these will prove so efficacious as to be profitable, and it is still more unlikely that native farmers would adopt their use.

In Nigeria it is found that twelve good pods, on the average, will yield about $2\frac{1}{2}$ lb. of wet beans or 1 lb. of dry cocoa: the percentage of dry cocoa to wet is commonly about 40 or 42 per cent. The number of pods per tree varies greatly according to the distance of the trees apart: in cocoa farms as closely planted as they commonly are in West Africa, the figure is usually something like twenty pods per tree. The yield per acre in West Africa is very high as compared to that in other countries. For instance on the five quite ordinary untreated native farmers' plots used in the experiments mentioned in the last paragraph, the highest yield recorded was 1400 lb. per acre and the average was 950 lb. per acre. In another series of five separate 1-acre plots, of which the yields were recorded some years ago, the average yield per plot per annum was from 1000 to 1300 lb. An ordinary average yield in the West Indies is apparently

2 or 3 cwt. per acre, and 4 or 5 cwt. is there an exceptionally good crop.

The Agricultural Departments of the Gold Coast and Nigeria have devoted their efforts, not so much to increasing production, as to improving the quality of the crop. Apart from differences that are inherent in different varieties of cocoa trees, quality in the commercial product depends upon the fermenting of the beans after harvesting, and upon their relative freedom from damage by mould or weevils. The pods are harvested by being cut from the tree, and are then split open with a cutlass and the beans extracted. The beans at this stage are embedded in a mass of mucilage. If the whole mass of beans and mucilage is kept in a heap, or in a box or basket, the mucilage ferments with an accompanying rise of temperature. During this process the sugar in the mucilage changes first to alcohol and then to acetic acid: and fermenting cocoa smells of these substances if the fermentation is proceeding properly. In West African cocoa, full fermentation takes six or seven days. Fully fermented beans, when dried, are chocolate-coloured inside, are friable, and have a flavour resembling that of unsweetened chocolate: under a powerful lens the colour can be seen to be uniformly distributed. The bean that is simply dried without being fermented is not friable, is of a dark slaty colour inside, and has no taste of chocolate: and under the lens the colour is seen to be restricted to small groups of purple cells on a white "ground". Partially fermented beans are of a purple colour and of a texture intermediate between fully fermented and slaty beans; and if they are examined with a lens the colour is seen to be still restricted to groups of cells. The manufacturers consider that both slaty and purple beans make cocoa powder or chocolate of inferior flavour as compared to that made from fermented beans. But the amount of importance attached to the fermentation of West African cocoa varies a good deal in different countries. This is due partly to the fact that the manufacturers of chocolate in different countries are accustomed to mix in with the West African cocoa different proportions of beans of the superior varieties, which are produced in the West Indies and in some of the South American countries. Moreover, the value of cocoa powder (for making the beverage) varies greatly in different countries. In the United States of America, for instance, the production of cocoa powder is already in excess of the demand, and West African cocoa is consequently generally regarded there simply as a source of cocoa butter for addition to the beans of the superior varieties used in the making of chocolates for eating. Thus the American manufacturer attaches far less importance to the proportions of fermented and unfermented beans in the West African cocoa that he buys, than does the manufacturer in England or on the Continent of Europe.

In order to ensure uniform fermentation, it is customary in other cocoa-growing countries to transfer the fermenting mass of cocoa each day during the process from one box into another so as to mix it thoroughly. Similarly it has been always recommended in the past that the method of fermenting in heaps, which is commonly used in West Africa, should be supplanted by the use of boxes or baskets to facilitate this thorough daily mixing: or at the very least that the heap should be

opened and remade daily. But experiments in Nigeria
show that the presence of unfermented beans in fermented cocoa is not necessarily, or even chiefly, due to
lack of sufficient turning and mixing. Moreover, the
experiments show that as good cocoa can be produced
by fermenting in a heap made up upon, and covered
with, banana leaves as can be prepared in boxes or
baskets. Nor does turning the mass daily or even twice
daily from one box to another, or daily opening and
remaking of the heap, result in a better final product
than is obtained by turning only twice in all, once after
two days and again after a further two days. Fermenting
in pits dug in the ground, however, gives an inferior
result to other methods. When fermenting only a
hundredweight or half-hundredweight of cocoa at a
time, as the West African farmer commonly does, the
heap is the most practical method.

The main cause of unfermented or under-fermented
beans in West African cocoa is the inclusion of some
unripe pods in the harvest. The facts about the fermentation of such cocoa have been found to be rather
peculiar. In some experiments made in Nigeria parcels
of exceedingly unripe cocoa were fermented;—much
less ripe than anyone would ever harvest in ordinary
practice. It is found that even in these circumstances a
proportion of the beans will change into fermented
cocoa in the ordinary way; but there is also a proportion,
which, however long the cocoa is fermented, or however
much the mass is turned, still remains slaty or purple.
Now it is not feasible to pick every pod in a field just
when it is exactly ripe, and, if the cocoa is left on the
trees too long, some of it becomes overripe and ger-

minates, or at least, when fermented for the usual six days, takes on the appearance which is commonly ascribed to fermentation for too long a time. In practice, therefore, it is not really possible to prepare an absolutely perfect quality of Nigerian cocoa, that is, one consisting of 100 per cent. of fermented beans, yet with none "under-fermented", "over-fermented" or germinated. But it is quite easy to prepare it so that none of the beans is slaty and only 3 or 4 per cent. purple. Such cocoa passes as grade 1 in Nigeria and easily passes as "Good fermented" (the highest grade) in Europe.

Much effort has been devoted in the Gold Coast and in Nigeria to encourage good fermenting by native farmers. Already in the Gold Coast, most of the cocoa is at least fairly well fermented; and in Nigeria over 30 per cent. of the cocoa produced in 1931–2 was fermented. Fermentation involves a certain small amount of trouble; and, moreover, a given quantity of fresh cocoa, if dried directly without being fermented at all, will yield 2 or 3 per cent. more dry cocoa by weight than it will if it is fermented before being dried. Naturally, therefore, the farmer will not ferment his crop unless he gets a higher price for fermented than for unfermented cocoa. Since the War, the Agricultural Department in Nigeria has devoted considerable effort to helping the producer of fermented cocoa to get for it the higher price that it is worth. This has been attempted in two ways; by grading all cocoa offered for sale, and by forming associations of producers of fermented cocoa and helping them to bulk their numerous small individual parcels, and to sell the whole co-operatively by tender. In the early years many difficulties had to be contended with; but the continued effort has begun to show its effect, and recently there has been a visible improvement in each year's crop as compared with that of the previous year.

Cocoa, whether fermented or not, remains free from mould provided only that it is dried reasonably quickly and thoroughly. Fortunately at the time when the main crop ripens in West Africa, in November, December and January, the weather in the Gold Coast, and still more in Nigeria, is so dry that cocoa only becomes mouldy through gross carelessness. The chief cause of mould in West African cocoa is packing it into sacks for transport while still damp. This has now been forbidden in Nigeria by a law which is stringently enforced, with the result that in the main season there are practically no mouldy beans at all in the cocoa exported; and not a very great amount even during the rainy months, when a little cocoa ripens "out-of-season". This state of affairs is in very marked contrast to that which existed six years ago.

It is hardly possible, either in Nigeria or in the Gold Coast, to dry the out-of-season cocoa in the rainy months sufficiently fast to prevent the occurrence of a small proportion of mouldy beans in the finished product. In the Cameroons the climate is so humid during the greater part of the year, that cocoa cannot be satisfactorily dried merely by spreading it out every day on a floor in the open. The estates are therefore equipped with special houses which are warmed by pipes which lead, in and out of the house, the hot gases and warm air produced by outside furnaces. Even the

small native farmers, under the guidance and encouragement of the Agricultural Department, are taking to
using simple drying floors warmed by fires placed under-
neath them. Dry cocoa, which is stored in a humid
atmosphere like that of the coast in the rains, will
reabsorb moisture and again become liable to mould.
Experiments in both the Gold Coast and Nigeria show
that it is only in the more humid parts of the cocoa belt
that storage in a good store is dangerous, but much
depends on the flooring and ventilation of the store.
The present practice of shipping the dry cocoa home
without delay is certainly the safer.

Damage by weevil only occurs to cocoa in West
Africa if it is stored in the country for some time before
shipment. If it is shipped reasonably quickly after purchase, there will be little or no infection by weevils, and
none at all by the more dangerous moth which attacks
stored products. The damage by weevil or moth that
sometimes occurs even to Nigerian cocoa seems to result
entirely from infection after the arrival of the cocoa
in England. Every bag of cocoa exported from Nigeria
is inspected by Government officials, after which the
bags are sealed with official seals which may not, except
by special authority, be removed in Nigeria. A similar
system was recently instituted in the Gold Coast.

KOLA (*COLA ACUMINATA*)

The superior species of kola tree grow wild in some
parts of the Gold Coast and Sierra Leone, and until
recently some 5000–8000 tons of nuts were exported
from these colonies annually to the adjacent West
African Colonies. Inferior varieties of kola grow wild,

or at least are indigenous, in Southern Nigeria; but the superior varieties were introduced perhaps thirty years ago. The production of kola nuts in Nigeria is steadily increasing, and the locally grown nuts are rapidly replacing those imported from the Gold Coast. The kola tree is grown entirely for the sake of its nuts, which the native of West Africa likes to chew. The nuts have a very astringent taste which is appreciated by few Europeans. They are said to have a great stimulating and sustaining effect, and to be a great help to the natives in the long marches which they will often make. These properties are due to the presence of the alkaloids, caffeine and theobromine. The kola tree is grown from seed, which must be fresh at the time of sowing. The trees grow to a height of 30 feet or so and require to be spaced at least 25 feet apart. They begin to bear when they are about seven years old, and the average annual yield increases steadily until they are about twelve years old. The annual yield is very variable indeed. The yield of one plot of kola trees on the Government Farm at Agege, near Lagos, has varied from 400 to 700 lb. of good nuts per acre, with an average of about 500 lb. per acre. At Ibadan the yields are lower; and the best yield recorded there has been 460 lb. per acre, while the average annual crop is only about 150 lb. per acre. The price of kola has been steadily falling for some years, and is now about 3s. per 100 nuts or about 1s. per pound for the best variety of nuts; thus, since this crop involves little labour, it is still a very profitable one.

The kola tree is much more hardy than the cocoa tree. Kola trees planted recently by the Agricultural Depart-

ment at Benin and at various places in the Eastern
Provinces of Nigeria, on soils where cocoa always fails,
have grown well, and are now beginning to bear fruit.
The local farmers are therefore beginning to plant
kola themselves, and there is the prospect of a growing
production of kola in these provinces in a few years'
time.

COCONUTS (*COCOS MICIFERA*)

The coconut is not greatly grown in West Africa. A
few trees are seen around most villages in the coastal
region of the Gold Coast, but their produce is generally
used only for local consumption. There are also a few
trees scattered throughout Southern Nigeria, especially
in the neighbourhood of Badagry. A small quantity of
copra is exported from the Gold Coast and efforts are
being made to stimulate the industry. The export of
copra from Nigeria is negligible.

RUBBER

There was at one time a considerable export of wild
rubber from West Africa, but in recent years this
has almost entirely ceased. There are also, in all the
West African Colonies, a few native-owned plantations
of Para rubber (*Heven Braziliensis*), but the plantation
rubber industry has never been developed in West
Africa and is unlikely to be in the future, in view of the
fact that the world's production already vastly exceeds
its needs.

CHAPTER 11

COTTON, GROUNDNUTS, BENNISEED, GINGER

COTTON (*GOSSYPIUM* Sp.)

IT is questionable whether cotton is strictly indigenous to the west coast of Africa or not; but it has certainly been grown by the natives for countless generations to supply local weaving industries, the products of some of which were famous long before the European arrived on the coast. These local industries still exist and absorb annually perhaps the equivalent of 10,000 bales of cotton or more in Nigeria alone. Several different species of indigenous cotton are grown in various parts of West Africa, and these are again subdivided into different strains which are especially suited to particular localities. Thus in Nigeria, the fuzzy seeded *Gossypium Peruvianum* or "Meko" cotton is mainly grown in the South-west Provinces; *Gossypium vitifolium* or "Ishan" cotton, which has a black naked seed, is mainly grown in the north of the Benin Province, the Kabba and Benue Provinces; and *Gossypium punctatum* was formerly grown in many parts of the Northern Provinces. These indigenous cottons are all characterized by having short, strong rough lint which, when spun and woven by hand, makes a very coarse cloth, though it is a cloth that will stand very long and hard wear. They are not very suitable for spinning in cotton mills, and consequently fetch a very low price on the world market; in fact when the world price of cotton is low, these kinds are almost unsaleable at any price. The value of cotton as an export crop has always been realized by the European in West Africa and many attempts have been made in all the British Colonies both to stimulate production, and to improve the quality. Until quite recently these attempts had met with very little success except in Northern Nigeria; for the problem of finding improved varieties suitable for export proved to be very difficult to solve. Even to-day it has not yet been accomplished either in the Northern Territories of the Gold Coast or in Sierra Leone, and cotton production in these areas has so far made no progress. In Northern Nigeria a solution was found by the introduction of an American Upland cotton—Allen's Longstaple—which proved able to adapt itself to the climatic conditions, and has now completely replaced the indigenous cottons in all the main cotton-growing areas.

Whether or not an even better type of cotton might in time have been evolved by scientific selection from the indigenous cotton, it is impossible to say. The introduction of an exotic variety proved to be an easy and rapid solution of the problem, and the Allen cotton at least has proved very satisfactory. Since its introduction, it has been steadily improved by selection; and the cotton which is now grown probably bears but a slight resemblance to the commercial American strain from which it originated. The limit of improvement, except in the direction of uniformity, seems to have been almost reached; and the standard already attained is so high that, although many other exotic types have subsequently been introduced and tested, none of them has proved to be capable of replacing the Allen variety. But even if no

further improvement can be effected, which is not yet
certain, the maintenance of the present standard necessitates continuous selection and multiplication of new
selections.

In Southern Nigeria exotic types of cotton, though
many were very thoroughly tested for several years, were
not successful. Cotton in Southern Nigeria is almost
invariably interplanted with other crops, and the exotic
varieties were unable to produce a good yield in competition with them. They also always suffered much
more severely than the indigenous cottons from insect
pests, and the fungus and virus diseases which those
pests spread. An attempt was therefore made to obtain
an improved strain by selection from among the native
varieties; and success has been attained with "Improved
Ishan" cotton. The only defect of this cotton is a
certain roughness of the lint, and it is hoped that by
further selection or crossing this fault may be eliminated. Agriculturally it is perfectly satisfactory; for it
yields well, is resistant to disease, and will succeed
in competition with other crops in a mixed field. In
Nigeria there is still a middle belt consisting of the
Ilorin, Kabba and Benue Provinces, where neither
American nor the improved Ishan cotton is satisfactory;
and efforts to evolve a suitable type for this area are still
being made. In developing the cotton industry, the
work of the Agricultural Department does not end with
the selection and multiplication of improved strains. In
order to secure the full premia for quality, the cotton
sold for export must be uniform, free from dirt or broken
leaves, and free from immature cotton. The seed cotton
therefore has to be inspected and graded in order to

induce the producer to market his crop in the best possible way. Also, since the seed cotton has to be taken to a ginnery in order to remove the seed from the lint, organization is required both to return seed to the producer and to see that he is supplied with seed of the right type. In Nigeria there are special regulations to ensure that all seed cotton for ginning for export is inspected and graded; and all such cotton has for this purpose to pass through special markets before it goes to the ginneries. These markets, though popular with sellers, have from time to time been criticized both by the trading firms and by local administrative officers; but that they have amply justified themselves is shown by the fact that the Nigerian cotton crop is marketed in a better condition than the crop from any other country in the world. It therefore enjoys a reputation for quality; which reputation has a money value that is passed on to the producer.

In Northern Nigeria, cotton has in the past almost invariably been grown as a sole crop; but, owing to the recent low price of cotton, some farmers are beginning to interplant their cotton in crops of "gero" or maize. These are early crops which can be harvested a few weeks after the cotton is planted. This practice may be regarded as an example of the way in which the native farmer can adapt his methods to suit economic conditions. By growing a corn crop as well as cotton on the same land, he is able to obtain a greater total return for his labour, even if he loses a little in his yield of cotton. He reduces to a minimum the cost of the labour actually expended on the cotton. To obtain a maximum yield in Nigeria, cotton seed should be sown in June in the north and early July in the south; but the sowing date for cotton is rather a gamble in the north. If it is sown too early it grows well, but is liable to suffer considerable damage from diseases and insect pests. If it is sown too late, then it is liable to have its growth prematurely checked by the early onset of the Harmattan. The farmer therefore has to decide between insect pests and the Harmattan. Actually, for other reasons, he usually risks the Harmattan, and much of the cotton in Northern Nigeria is sown far too late to give a maximum yield. The farmer's first consideration is his corn crop, so that the cultivation of land for cotton usually has to wait until all the weeding of guinea corn is finished. He also often plants cotton on land which is being newly broken for guinea corn in the following season; and such cotton is inevitably sown very late indeed. But his attitude towards such a cotton crop is that even a small yield is better than nothing; and cotton is the only crop that can be grown in these circumstances. In any event, the land receives a thorough digging, and the cotton seed is planted on freshly made ridges. In its early stages, cotton suffers more severely than almost any other crop from the competition of weeds, so that it responds greatly to frequent cultivation and weeding; and the farmer makes every effort to supply this. Cotton will grow on almost any soil; but in Northern Nigeria it is usually grown on the heavier land, which is not very suitable for groundnuts. Cotton has a tap root which can penetrate the soil to a great depth. It therefore does not do well on shallow soils; but on deep and heavy soils it will continue growing long into the dry season. The leaves are often still green after the cotton picking is finished, and are eventually eaten by cattle and goats. The normal seed rate in Northern Nigeria is about 20 lb. per acre, and the plants are thinned to two or three at a stand when they are 6–9 inches high. The bolls begin to open in November and picking continues until the end of January. The yield of cotton in the north is largely influenced by climatic factors. One or two storms in the early part of October, totalling 2 inches or so of rain, cause big yields, whereas the onset of the Harmattan a week or two earlier than usual causes a low yield and the occurrence of a high proportion of immature lint. It is doubtful whether the average yield on native farms exceeds 120–150 lb. of seed cotton per acre; but on well-manured land, yields of 450 lb. per acre may be obtained if the season is favourable. On the Government experimental farms, 300 lb. of seed cotton per acre is considered to be a good average yield.

In Southern Nigeria cotton does not receive much special cultivation. It is almost invariably interplanted with either yams or maize, the seed being sown on the sides of the mounds in clumps. The young plants have more or less to look after themselves. The land is weeded only if the yam or maize crop needs it. Generally the cotton plants are not thinned out, so that there may at first be five or six plants at each stand, of which two or three will grow into mature plants. The seed may be sown at any time from July to September, but the best yields will only be obtained if the seed is sown early in July. Cotton in the south of Nigeria, unlike that in the north, suffers very greatly from insect pests and disease, especially from the insect "Cotton Stainer" (*Dysdercus* sp.). But in spite of this, under favourable conditions, much higher yields can be obtained in the south than in the north. On the Government farm at Ibadan a yield of over 1000 lb. of seed cotton per acre has been recorded, while yields of 500 lb. of seed cotton per acre are normally expected. Harvesting takes place from January to March, but the bulk of the crop is ready for ginning in February.

GROUNDNUTS (*ARACHIS HYPOGAEA*)

British and French West Africa taken together constituted, until a year or two ago, the greatest groundnut-producing area in the world. The export of groundnuts from British India is now greater than that from West Africa; but the latter still exports some 500,000 tons annually. In the British Colonies the industry is confined almost entirely to the Gambia and to Northern Nigeria. The Gambia almost depends for its existence on the groundnut, and its cultivation has long been established there. In Northern Nigeria the development of the industry on a big scale is comparatively recent, and dates from the completion of the railway to Kano. Groundnuts will grow almost anywhere in West Africa; but since, in normal times, the world price per ton is comparatively low, the industry cannot develop in any particular district until it has been provided with adequate means of transport; for the native will not carry groundnuts in head loads for excessive distances. If the means of transport are inadequate, he will prefer to grow cotton. A good deal of the soil in the Northern Province of the Gold Coast is suitable for groundnuts, but in the absence of a railway the industry cannot develop. Generally speaking, cotton and groundnuts do not com-

pete to a very great extent; for cotton is a crop for heavy
land and the groundnut is a crop for light sandy soils.
One of the main problems of the plant breeder in
Northern Nigeria is to evolve a strain of groundnut
which will thrive and yield well on heavy soils, prefer-
ably an upright type which can be lifted by means of
cattle-drawn implements. If groundnuts are grown on
heavy soils, their harvesting is a most laborious operation; for in the very dry climate of Northern Nigeria
such soils harden suddenly and completely as soon as
the rains cease, when the groundnuts are still barely
ripe. Thus measures which would effectively reduce the
labour of harvesting on heavy soils would be a real improvement, and might be expected to find favour with
the farmer.

As with most other crops in the tropics, there are
many local varieties of groundnut, each of which is
particularly adapted to its own district. The chief difference in variety is that between the spreading and the
erect types. The erect types are more suited to interplanting with corn crops than the spreading types;
otherwise, the latter generally produce the greater yield.
Several exotic varieties have been tested in Nigeria; but
all have been found to yield less than the local types,
especially on heavy soils. Thus it seems as if selection
and multiplication of superior plants in the indigenous
types will in the end be most likely to lead to improved
strains.

Groundnuts are more usually grown as a sole crop than
in mixed culture; but they are very often interplanted with
a few rows of gero. The gero, no matter how thinly it is
planted, is found by experiment to have a great adverse

effect on the yield of the groundnuts; but the profit or loss of the practice of mixing the crops depends upon the relative prices of corn and groundnuts at harvest time, and thus cannot be foretold at the time of planting. The groundnuts are usually planted either on the flat or on small ridges about 3 feet apart; and in the extreme north, with its short wet season, the seed is sown as soon as possible after the first rains arrive. Further south, at Zaria, where the rains last longer, groundnuts are not planted till June, as they require only about five months in the ground, and dry weather at harvest time is essential. Still further south, at Ibadan, where there is a break in the rains during August, it is possible to harvest the nuts during this break; and in these areas the earlier the seed is planted the better, for late-sown groundnuts suffer severely from rosette disease. At Ilorin it is better to sow the nuts in the shell before the rains break rather than at the end of April. This is a point which the agricultural officer has been able to teach the native farmer. His normal practice there was to plant in June or July and harvest the nuts in November. The crop therefore suffered badly from rosette disease and the yield was very small indeed. The seed may be planted either shelled or unshelled at a rate equivalent to 12 to 20 lb. per acre of shelled kernels.

Groundnuts require careful weeding in the early stages, but the spreading types soon completely cover the ground and thereafter require little further weeding. The soil around the plants needs to be loose to enable the plant to bury the nuts. The flowers are produced above ground, and, as soon as the seed has been set, the flower stem elongates and turns downwards forcing the young fruit below the surface of the ground; and it will not mature unless it can get into the soil.

Harvesting has normally to be done by hand. The plants are pulled up with the nuts adhering to them and are laid roots upwards on the tops of the ridges for a day or two to allow the nuts to dry. These are then picked off, and the rest of the plant is stacked for fodder for horses and cattle, and is highly valued for this purpose. After this, the ground is lightly hoed to recover any nuts which may have been left in the ground. On the experimental farms in the south, where there are no cattle, the haulms are buried after the removal of the nuts. If this is done, the groundnut crop will actually increase the fertility of the soil for succeeding crops, in spite of the removal of a heavy crop of nuts.

On suitable soils a yield of 1500–2000 lb. of undecorticated nuts per acre is quite normal, and even higher yields have from time to time been recorded on light land. On heavy and wet soils, such as those of Zaria, a yield of 800–1000 lb. of undecorticated nuts per acre is regarded as a good yield, while 500–600 lb. per acre is an average yield.

Before groundnuts are marketed for export from Nigeria, the kernels are separated from the shells. This is not a very laborious process, although it is carried out by primitive methods. In some districts the nuts are beaten with sticks, but more commonly they are pounded with a wooden pestle in a wooden mortar. Either process results in a large proportion of broken kernels, which slightly lowers the value of the produce. Certain firms are investigating the possibility of using machinery for this purpose; but so far little progress has been made in this direction. The proportion of kernel to nut usually approximates to 70 per cent., but the figure varies somewhat for different varieties. Thus in a test carried out at Kano in 1930 it was found that while "Damberta" or local nuts gave a hulling percentage of 74 per cent., "Phillipine White" gave 67·5 per cent., and a rather peculiar variety from Biu gave only 48 per cent. In addition to the export trade, there is also a small local trade in both groundnuts and groundnut oil. The oil is used for cooking purposes by both natives and Europeans.

Groundnuts are pressed in Europe to extract the oil. The oil content being commonly 48 or 49 per cent. of the weight of the decorticated kernels. The oil is used in margarine making to some extent, but its chief use is as a cooking or salad oil, in substitution for olive oil. Marseilles is the most important port and manufacturing centre in the world's groundnut trade, though a considerable quantity is annually pressed at Liverpool and Hull. The cake remaining after the expression of the oil is a valuable cattle food. It is, however, not commonly used by the British farmer in the pure state, but constitutes an important part of the proprietary brands of compound cakes and meals.

BENNISEED (*SESAMUM INDICUM*)

Benniseed is grown on a very small scale in many parts of West Africa and particularly in Northern Nigeria; but as a rule it is merely one of the many minor crops grown in the compound around the native farmer's hut on a garden scale. Among the Munshis of the Benue Province of Nigeria, during comparatively recent years, there has, however, been some development of benniseed production. It is now the main export crop of this particular tribe, and the annual export is increasing slowly but steadily.

The quantity exported annually, some 2000–3000 tons, is still comparatively small; but this trade is interesting as an example of the way in which the native can develop an industry entirely on his own, almost in spite of the European. The Munshis have apparently always grown a little benniseed; but the development of the industry commenced with the introduction of taxation and the necessity of finding cash with which to pay the taxes. At that time the Munshi farmer had a choice between cotton and benniseed as his money crop. He deliberately chose benniseed in spite of repeated efforts to induce him to develop the cotton-growing industry. He even went so far as to alter his traditional rotation of crops in order to include benniseed as a sole crop. There is no doubt whatever that he made the right choice. Benniseed paid him better than cotton.

As cultivated by the Munshis, benniseed can be grown either as an early or as a late crop. Seed for the early crop is broadcasted in April. The land is usually roughly hoed and weeded before the seed is sown, and the crop may get one subsequent weeding, or it may receive no further cultivation at all. The early crop is harvested in August, and an average yield from native farms is about 200–250 lb. of seed per acre. On the Government experimental farms, however, 300 lb. of seed per acre is regarded as a normal yield. Late benniseed is sown in the latter part of August at the end of the break in the rains, and is harvested in November or December. The cultivation is similar to that of early benniseed, but the yield is considerably less. Benniseed suffers little from insect pests or fungus diseases.

The Munshis' method of harvesting benniseed is somewhat laborious. The crop, after being cut, is tied into small bundles which are then hung from wooden rails which have to be specially erected for the purpose; the bundles remaining on the rails until the seed pods are dry when the seed can be easily beaten out. The drying crop needs special care, for the seed sheds readily as soon as the pods are dry. Repeated and careful experiments have shown that the European method of stooking corn can be used for benniseed with a very considerable saving of labour and no appreciable loss of seed. This method has therefore been demonstrated to the Munshi farmers recently, and has already been adopted by a few of them.

In Nigeria the commercial value of benniseed is commonly lowered by the presence in the crops of a proportion of two other plants—*Sesamum Radiatum* and *Ceratotheca Sesamoides*. These are both closely related to the true benniseed, and bear very similar seeds which contain oil somewhat similar to that of the true benniseed. But whereas the true benniseed contains 50 per cent. of oil or rather more, the seeds of the adulterants contain only 32–37 per cent. The seed of the true benniseed is white, light yellow or pale brown in colour, while the seeds of *Sesamum Radiatum* are always black or very dark brown, so that their presence is easily detected in a sample of benniseed. *Ceratotheca Sesamoides* produces seeds of rather varied colour, black, dark grey, brown or yellow; the yellow seeds of this species are not very readily distinguishable from those of the true benniseed. But, since even the *Ceratotheca Sesamoides* always includes some plants yielding dark seeds, it will be found in practice that a sample which is free from dark seeds will consist solely of true benniseed. If any dark seeds are present the sample is not pure, though the proportion of dark seeds does not always represent exactly the true percentage of impurity.

The inferior plants are more easily distinguished in the field than is the seed in the harvested crop, as the flowers of the true benniseed are always white, while those of the other two species are purple. The trading firms and the Agricultural Department in co-operation have issued large quantities of pure seed which was given to growers in exchange for an equal quantity of impure seed. By this means the crop of a large part of the Benue Province has already been brought into a pure state, and it is hoped that before long the whole of the area will be similarly treated.

Most of the benniseed from Nigeria is exported to the Continent. In some continental countries the admixture of a certain proportion of benniseed oil in all margarine is enforced by law, in order to facilitate the detection of margarine as an adulterant in butter. For it happens that there is a peculiarly sensitive chemical test for benniseed oil by which its presence in even small proportions is readily detected. In India and Turkey benniseed is the most highly valued of all edible oils, and is especially used in making sweetmeats. The price which benniseed fetches in Europe seems sometimes to be hardly as high as the price in Turkey.

The price paid in Nigeria seems always to leave a greater margin of profit for the exporter than is obtained on other Nigerian exports. Recently the local prices of benniseed and groundnuts have been much the same; but on the average of the last ten years benniseed has been bought locally at a lower price than groundnuts, although it fetches a higher price in England. Its yield per acre is only about a third of that of groundnuts or even less, so that the return to the grower in money per acre from benniseed is considerably less than from groundnuts, and must always be so unless its local value per ton is much higher than that of groundnuts.

Benniseed is generally regarded as a crop that exhausts the fertility of the soil more than most crops, but experiments carried out in Nigeria have not confirmed this belief.

GINGER (*ZINGIBER OFFICINALE*)

Ginger is one of the chief export crops of Sierra Leone, and it is also grown in very small quantities in several other parts of West Africa for local consumption as a medicine. Recently, efforts have been made in Nigeria to stimulate its production among the pagans of the Zaria Province. These efforts have had encouraging results, and ginger cultivation is now beginning to spread to other provinces.

Owing to its comparatively high price per pound, ginger is a suitable crop for districts remote from railways and motor roads, where it does not pay to grow crops such as cotton and groundnuts because of the high cost of transport. But from the farmer's point of view, it has the disadvantage of requiring a tedious preparation of peeling, washing and drying before it is ready for the market. This is not such a serious objection where the work is done by the women, as is the case among the pagan tribes in Zaria. In Sierra Leone the ginger is not peeled, and in consequence fetches a very low price when sold in London.

The ginger of commerce is the rhizome or thickened underground stem of the ginger plant. These rhizomes, owing to their shape, are usually referred to as "hands". The cultivation of ginger is quite simple. It requires a light well-drained soil and plenty of manure or organic matter. In the absence of manure, the Zaria pagan farmers usually grow their ginger in the wooded valleys where the soil contains leaf mould derived from the leaves of the trees; but the shade of the trees seems to tend to make the ginger fibrous. Thus if manure is available, it is preferable to grow the ginger in the open. The valley fields are very small indeed, seldom exceeding an area of $\frac{1}{16}$ acre each. The "seed" consists of short pieces of rhizome saved from the previous year's crop. It is sown in April or May, usually on the flat, and the crop is harvested in the following dry season. The only attention which is normally given to it by the Zaria pagans is to pull out the weeds by hand; but ginger readily responds to cultivation with the hoe, and efforts are now being made to induce the farmers to plant their ginger on ridges and to give it such cultivation.

When harvested, the hands are covered with a thick peel, and the removal of this peel is the essential part of the preparation of the crop for the market. There are several slightly different methods of preparing ginger, some of which involve scalding; but whatever method is adopted the object is the same, namely, to remove the outer skin without damaging the hand. The criteria by which a sample of ginger is judged, are the relative plumpness of the hands, external whiteness, and softness and lack of fibre when cut. Of these, whiteness can only be obtained by careful peeling, washing, and drying. The first stage of the method adopted in Nigeria is thorough washing before peeling, and the removal of all sand and roots. Then the ginger is allowed to soak overnight in more clean water, after which it is scraped with blunt knives. After scraping it is washed several times, and then allowed to dry in the sun. The whole process takes about nine days. A woman can peel about 20 lb. of ginger per day.

The yield of ginger is anything up to 6 tons per acre of fresh ginger, and fresh ginger yields approximately one-sixth of its weight of dry peeled ginger. The price paid in Nigeria for cured ginger is commonly about $2\frac{1}{2}d.$ per pound, or about £23 per ton. Nigerian ginger prepared by the above method fetches about the same price in London as Jamaican No. 3 ginger; and that price fluctuates between £35 and £70 per ton. There is, however, no reason why the quality of Nigerian ginger should not be further improved; and if the industry is to develop, quality is all important. The chief defect at present is the small size and fibrous texture of the hands. The colour and peeling are satisfactory, and the pungency is equal to that of Jamaican ginger, so that there is nothing inherently wrong with the quality of the Nigerian product. The plumpness of the hands can be improved by selection and better cultivation. This aspect of the industry is now being studied. Although it will probably never be grown on more than a garden scale, yet it is a very useful crop for the small man, and, provided he will take the trouble to produce a high quality, it is very profitable.

CHAPTER 12

CEREAL CROPS

MAIZE (*ZEA MAYS*)

MAIZE is grown to some extent everywhere in Nigeria and the Gold Coast except in the extreme north; in the districts within 90 or 100 miles of the coast it is the only cereal crop. There are several local varieties, which differ chiefly in the length of their growing period. It was presumably introduced several centuries ago, but many attempts that have since been made to introduce new exotic varieties have met with no success, as the native finds them less palatable than those to which he is accustomed.

The early maize crop in the south corresponds to the gero crop of the north, in that it is planted with the first rains and is the first food crop of the year to come on to the market. If planted in March or early April it is ripe in August; but much of the crop is harvested while still soft. By eating it in this stage the grower obtains a much more palatable food, and his wife avoids the rather laborious process of reducing the hard ripe grain to flour. In favoured positions, as on the banks of streams, maize can be, and is, planted in nearly every month of the year.

Early maize is usually planted on the sides of stale yam heaps without any digging or turning of the soil, and receives no cultivation other than weeding. The seed is sown at the rate of about 16 lb. per acre. Generally cotton, guinea corn or beans are planted through it at any time after it is well established.

If no other crop is interplanted in it, and the maize is sown correspondingly thick, an average good field of early maize will yield about 2000 lb. of dry grain per acre, but the yield obtained by the native farmer is not easily ascertainable. It is probably little more than 1000 lb. per acre, for his crop of maize is obviously much smaller than those on the experimental farms.

In Southern Nigeria, late maize is usually planted as soon as the rains recommence at the end of August, and it is harvested in December or January. It is often planted on the flat on newly cleared land where it is intended to plant yams in the following season. The land is cleared in slack periods during the early rains, and, if time permits, the yam heaps are made before the maize is planted. The yield of late maize is at best only about half that of the early crop, and it is liable to rather more seasonal variation than the early crop—which latter varies remarkably little from year to year. Shortage of rain is the only factor which will seriously affect the yield of maize in West Africa, provided that it is grown on good soil: for maize is particularly free from disease or from insect pests. For this reason, and because it responds greatly to any differences in the fertility of the soil, it is most useful on an experimental farm as an index of fertility.

Previous to the War there was a small annual export of maize to Europe, but from the end of the War until 1930 the price of maize in Nigeria, £4 per ton or even a little more, was too high to permit of profitable export. In 1931 the price fell to as low as £2 per ton at harvest time in the local markets, while it was about £5 per ton in Liverpool. If this margin should continue, export might be possible again if maize were accorded especially low freights by rail and steamer, as it is in other maize-exporting countries. If an export were to be established, it would be necessary for the grain to be passed through a "conditioning" plant to kill the weevils by which it becomes infested soon after harvest. Moreover, as the grower finds considerable difficulty in completely drying his grain, it would probably pay an exporter to do this artificially. The plant for both these purposes is not very expensive either in capital or running cost. There is even less opportunity for an export trade from the Gold Coast owing to the high price which maize fetches there.

GUINEA CORN (*SORGHUM VULGARE*)

Guinea corn is the main food crop of the north. Although it can sometimes be grown successfully as far south as Ibadan the climate there is really a little too wet for guinea corn. Thus its cultivation in Nigeria is, in practice, confined almost entirely to the Northern Provinces. The northward limit of its cultivation depends on the length of the rainy season, as it requires between seven and eight months on the land, in at least five of which the rainfall must be sufficient to keep it growing steadily. Right on the northern boundary of Nigeria the rainy season is frequently too short for guinea corn to succeed.

In those areas where it can be cultivated, guinea corn is literally the "staff of life", and every part of the plant is made use of in some way or another. The flour is used for human consumption; the bran or "dusa" is used for feeding to horses and cattle; the dried leaves, stripped from the stalks, are used as fodder; and the stalks are

used in the construction of houses and fences, and also
for fuel. Even the chaff has its uses; for it is either fed
to cattle or goats, or burnt in order that the ashes may be
used as manure. In some areas, for example around
Kano, the stalks have a very definite money value, and
are bought and sold in the market. Guinea corn is to
the farmer more than his main foodstuff. It is also a
money crop, and there is a considerable wholesale trade
in this grain. In Northern Nigeria the population of the
tin fields absorbs large quantities of grain, as do also the
populations, human and equine, of the towns. The farmer
who is fortunately situated near to a market or means
of transport therefore cultivates more guinea corn than
he requires for his own use in order to meet this demand.

The varieties of guinea corn are very numerous, indeed
almost every district has its own, which is specially
adapted to the local conditions. The varieties in Nigeria,
as in other countries where this grain is grown, are
peculiarly highly localized. Thus, although the botanists
of the Agricultural Department at Zaria have recently
selected varieties which in that locality yield much more
heavily than the ordinary farmers' types, it cannot be
expected that any particular improved strain will maintain its superiority outside a radius of 30 or 40 miles
from the breeding station. There are two main types of
Nigerian guinea corn, a type with a compact upright
head and a type with a loose open head. The loose-
headed type is the more common of the two; the compact type is confined to the drier areas.

Guinea corn can be, and is, grown on almost any type
of soil; but it does better on heavier than on lighter
soils, unless the latter are well manured. On either,

however, it responds very greatly to manure, and the native farmer consequently sows guinea corn on any land on which cattle have been kraaled during the dry season.

The seed, at the rate of 15–20 lb. per acre, is planted as soon as possible after the rains have begun, either on the flat or on old stale ridges. The northern farmer has very little time for cultivation before the seed of his cereal crops is sown; he must take advantage of the early rains and cannot wait until the soil becomes soft enough for deep cultivation. This comes later, while the crop is growing. The final ridging, before the dry season sets in, is very carefully done and the land remains clean until the next young crop is big enough to bear weeding. Where new land is being brought into cultivation, or old fallows are being re-cultivated, a common practice is to ridge the land in July or August of the previous season, take a crop of cotton, and then plant guinea corn on those ridges when the rain comes. The object of these practices is to take advantage of the dry weather in killing weeds which are upturned by ridging just at the end of the rains.

Guinea corn is often planted as a sole crop, but equally often it is interplanted in gero (bulrush millet). Planted among gero it does little more than establish itself while the gero is on the land; but it grows at an astonishing rate as soon as the gero is harvested in early August. This system insures the farmer against the failure of one of the crops in a country where the rains, especially the early ones, are uncertain. But if the early rains are good and the gero crop does particularly well, the guinea corn suffers accordingly.

As soon as the plants of guinea corn are about 6 inches high, i.e. 14–21 days after planting, they are weeded and thinned to two or three plants per stand. Blanks are usually filled up by transplanting, and if this is carefully done, the transplants seem to suffer very little ill effect from the operation. After this nothing more is done except one weeding until the gero is harvested in August when the crop gets a thorough cultivation, and the ridges are then made up with soil taken from the furrows. The general custom therefore is to give guinea corn altogether three weedings, each of which has a special name.

The crop is harvested towards the end of November or early in December. The stalks, which are then 10–12 feet high, are first cut and laid in the furrows; then the heads are cut off and made into neat bundles of about 60 lb. each, in which form it is normally stored and marketed. The grain is threshed by beating the heads with a stick and the proportion of grain to the remainder of the head (stalk, glumes, etc.) is 70–75 per cent. On manured land guinea corn will produce 1200 to 1300 lb. of grain on the head per acre, but the average yield obtained by the native farmer is only about 600–800 lb. on the head.

Guinea corn suffers very little from disease or the attacks of insect pests. Smut is present, but does very little damage: in certain seasons some slight damage is also done by a stalk-boring insect. The most serious pest of guinea corn is a root parasite, *Striga Senegalensis*, which, if not weeded out of a field, may in two or three seasons make the cultivation of cereals there almost impossible. Fortunately, owing to the fact that it produces a bright purple flower which can easily be seen, weeding it out is not difficult, and by continuously doing so before it sets seed it can be, and generally is, effectively controlled.

GERO or BULRUSH MILLET
(*PENNISETUM TYPHOIDEUM*)

Gero is the early corn crop of the drier parts of West Africa. In the central belts of the Northern Provinces of Northern Nigeria, gero, though often grown alone, is most commonly merely a catch crop in guinea corn fields. In this mixture the gero is sown first; but it is more thinly planted than the guinea corn and occupies the land for a shorter time, so that the latter is the more important constituent of the mixture. Further south in Ilorin, the place of gero in this mixture is frequently taken by maize. On the other hand along the northern boundary of Nigeria, where the rainfall is only some 25 inches per annum or less, and where the soil is commonly very sandy, guinea corn becomes a speculative crop, that will succeed in some years but fail more often; and there gero is the only cereal that is grown on a considerable scale.

Gero is very adaptable as to soil, and good yields can be obtained on almost any type, but it does best on well-drained light soils. It will not grow well on newly cultivated bush land. Gero is planted with the very first shower of rain—in some districts the seed may even be put into the ground before the rain actually comes, when it appears to be imminent. The seed rate is from 8 to 10 lb. per acre. It often receives no preparatory cultivation at all: at most a small heap of soil is made in which the seed is sown. In the extreme north it is often grown on the flat; but in the area where other crops are grown on ridges, so that the land in which the gero is sown is already ridged, small heaps for the gero are usually made in the furrows between the old ridges. Once the seed has germinated, young gero plants can stand quite long periods of drought. It tillers very freely, and within remarkably wide limits; its yield is not appreciably affected by the number of stands per acre. It generally receives no subsequent cultivation beyond a shallow weeding.

The grain is ready for harvesting about 4½ months after the seed is sown, and is usually stored on the head. The average yield is from 800 to 1000 lb. per acre on the head, or 400–500 lb. per acre of threshed grain. The threshing percentage is about 55 per cent. In the driest parts of Nigeria, gero is usually interplanted with cowpeas which are left on the ground when the gero is harvested; but in those areas where the wet season continues till the end of September or early October it is often followed by cotton, cassava or sweet potatoes. The introduction of mixed farming with its almost unlimited supply of labour and adequate supplies of manure should enable such intensification of the cropping to become more general.

MAIWA (*PENNISETUM SPICATUM*)

"Maiwa" is a form of millet closely resembling gero, but is less highly esteemed as a foodstuff. It is planted at the commencement of the rains, but it takes much longer to mature than gero, and is only ready for harvesting during the dry season at the same time as guinea corn. Its yield is rather less than that of gero, from 400–600 lb. per acre on the head.

MINOR GRAIN CROPS

In Northern Nigeria "small millets" are commonly cultivated for grain. They are hardy, and although they have no great agricultural importance, they form a useful supplement to the main grain crops, especially among the pagan tribes. The most important of these are "Acha" (*Digitaria exilis*), "Iburu" (*Digitaria Ibura*) and "Tamba" (*Eleusine Corocana*). The latter is an unimproved form of the Finger Millet which is grown in India and Ceylon. Iburu and Acha are extensively grown by the pagans of the Plateau and Zaria Provinces of Nigeria. They are usually sown broadcast on the flat, and yield from 600–800 lb. of grain per acre.

Tamba has a bigger grain than either Acha or Iburu, receives more careful cultivation, and is usually grown on ridges. As it is commonly interplanted with other crops, it is difficult to give a figure for an average yield, but it is in the neighbourhood of 600 lb. per acre.

RICE (*ORYZA SATIVA*)

Rice is more widely grown in West Africa than is generally realized; and its cultivation has increased in recent years. In Sierra Leone rice growing is well established, and rice is one of the most important foodstuffs of the country. In Nigeria and the Gold Coast the Governments have attempted to stimulate the industry by the erection of experimental rice mills. But on the whole, the native of West Africa is not a rice eater in the sense that the Indian or the Burman is; and rice, to the great bulk of the population, is still something of a luxury.

There are many varieties of local rice in Nigeria; but they all have a grain with a red skin, which makes them unsuitable for export to European countries, even when they have been cleaned by machinery. They also shed their grain very easily, and much of the crop is wasted through this cause. They are, however, very adaptable to varying conditions, and are able to withstand very rapid changes of water depth. Attempts are being made in Nigeria to introduce exotic varieties of white rice, but so far these attempts have not been entirely successful. The improved rices are less adaptable than the local varieties, and are more difficult to establish, so that unless thay are grown under the most favourable conditions they yield less than the more hardy native types. They are much less prone to shed their grain, and they of course produce a much better sample of milled rice. This latter point is of little interest to the native farmer who is only growing rice for local consumption, but becomes of great importance when a mill has been erected and an export trade to other areas is developed. Even if suitable exotic rices are eventually found, it will be a very difficult matter to keep them even reasonably pure, owing to the presence of wild and volunteer rice in all the rice fields.

Rice is grown in West Africa under both "hill" and "swamp" conditions; but in either case the West African is not an expert rice grower, and his yields are invariably very low. "Hill" rice is usually planted in the spaces between the mounds or ridges made for other crops. The seed is sown in clumps and is germinated by the rain. It receives no special cultivation, and depends on rain or wet-season springs to keep the land wet enough to enable it to grow. The common terms for this kind of rice—"hill rice" or "dry-land rice"—are deceptive, for a good yield, even of these varieties, is only obtained on very wet soil. But the conditions in which it is grown are none the less in sharp contrast with those in which "swamp rice" is grown, i.e. actually standing in water constantly. The growing of hill rice seems to be spreading in Nigeria.

Swamp rice is grown in the valleys of the larger streams and depends for its water supply on their annual floods. The floods sometimes occur rather suddenly, so that the rice crop may one day be standing in an inch of water and a few days later in 3 or 4 feet. The height of the annual flood also varies from year to year; and if the flood is smaller than usual, it will not reach the higher rice fields at all. Land for swamp rice is usually prepared during the dry season by roughly breaking it up into large clods which are left to bake in the sun. Puddling is not practised in Nigeria. The seed is usually broadcasted on the clods, when they are breaking down under the influence of rain and flood. The seed is germinated by rain, and the farmer endeavours to sow at such a time that the young plants can take advantage of the gradual rise of a normal flood. As the flood recedes and the fields dry out, the grain ripens and is harvested.

The yield of rice varies greatly. On our own experimental farms, yields as high as 2800–3000 lb. of paddy (rice in husk) per acre have been recorded. An average yield on a native farmer's field would appear to be from 600 to 800 lb. of paddy per acre. The paddy yields about

60 per cent. of cleaned rice. The yield of rice grown under hill conditions is very much lower than that of swamp rice.

There is a good deal that the European can teach the native farmer about rice growing; and if better methods of cultivation and better varieties can be introduced, the industry is capable of considerable development in West Africa. This has been shown to be true at Sokoto, where, by controlling the flood by means of a bank, by puddling the land and by transplanting seedlings from a nursery instead of broadcasting the seed, yields of as much as 2 tons (paddy) per acre of white exotic rice have been obtained, which is a far bigger yield than the native farmer can ever hope to obtain by his present methods.

CHAPTER 13

ROOT CROPS, BEANS AND MINOR CROPS

YAMS (*DIOSCOREA* Sp.)

THE yam is the main food crop of Southern Nigeria, and is an important food in the coastal belts of the other colonies. In Southern Nigeria one might almost say that all other crops are subsidiary to it. Deep cultivation, and the best land, are always reserved for this crop. The commencement of the yam harvest is an important event in the social life of the native, and the actual day is often formally fixed by the chief of the tribe. The varieties of yams are innumerable, most districts having a special variety which is locally much more popular than any other; but, in the main, the differences between the varieties are not very great. Some are earlier than others, some have whiter flesh, some are more suitable for pounding than others, and so on. The most important of these characters is that of earliness or lateness. Where yams are interplanted with cotton, or where they are grown in a rotation which incorporates green manuring, there is obviously a great advantage in growing a variety which can be harvested as early as possible.

The yam crop often occupies the land for nearly a whole year. The best time for planting varies somewhat in different localities, but generally the crop can be planted either in November or early December, or in March or April. Yams are not often planted in the interval between these two periods, and experiments indicate that this is not merely a question of convenience, as, if they are planted then, they do not germinate well. Also, yams planted too early do not do so well as those planted in November. The choice between the two good periods mainly depends on the proportion of land which can be prepared at the end of the rains; for yams planted in the earlier period yield better than those planted in March.

The farmer starts working on his yam land just before the rains cease and continues until the hardness of the soil makes the work too laborious. Such areas are planted in November–December. If he then has insufficient land prepared, he waits until the rains break when the ground must be prepared and the yams planted as quickly as possible. Any further delay results in a very serious loss of yield. In areas where the yam beetle (*Heteroligus Claudius*) is prevalent, late planting lessens the damage done by that insect. In spite of this, however, late planting usually results in a loss of yield, even when the beetle is prevalent; in other words late planting reduces the yield more than the beetle does.

The farmer almost invariably plants his yams on hills and the land then receives a very thorough cultivation. The hills are commonly about 2 feet high and are from 3 to 4 feet apart. The making of these hills involves an immense amount of labour, as there are some 1700 to 2000 hills per acre. In swampy land, or where especially large yams are required, the hills may be as much as 5 feet high, and on swampy land it is not uncommon to see rice growing between the yam heaps. When the land is cleared, the stems of the bushes and young trees are usually left and serve as supports for the yam vines. These are often supplemented by stakes; for the farmer fully realizes that staking results in an increased yield, as is easily demonstrated by experiment. In some parts, for lack of stakes, the stalks of the previous year's guinea corn crop are used as supports for the vines. Experiments have shown that the bigger the hill or ridge and the deeper the cultivation, the better the yield. This of course applies to some other crops; but the yam responds much more than most crops to deep and thorough cultivation of the soil before planting.

Any part of the tuber can be used as "seed", but the best germination is obtained by planting either a piece of the head of a large yam or a small whole yam. Pieces cut from the middle and bottom of cut yams germinate very irregularly and result in many blanks. The buds, unlike those on a potato, are not easily visible, and a section of the middle or the bottom of a yam may include no live bud. The crop requires, and repays, frequent weeding from the time of planting almost until it is ready for harvest. It is rarely allowed to grow as a sole crop, being usually interplanted with maize, beans, cotton or gourds.

The usual method of harvesting is to remove the tuber from the hill, without disturbing the vine, a few weeks before the latter dies. This results in the secondary growth of several small tubers which, when they are subsequently harvested, are used as seed for the next crop. If this is not done, the yams, in some districts, are left in the ground until they are required for food, obviating the necessity for special storage. Where theft is to be feared, they are stored by carefully tying them up on wooden racks, which are shaded, but exposed to the air.

It is probable that the yield of yams on native farms rarely exceeds 3 tons per acre; but on an experimental farm yields of 5 and even 6 tons per acre are not uncommon with some varieties. In many districts the yam, in addition to providing the farmers' food, is also a money crop; for in both Nigeria and the Gold Coast there is quite a large internal trade in yams. In the Gold Coast they are sent from the more northern areas to feed the population of the cocoa belt and of the coast towns; and in Nigeria they are exported both to the north and to the south from such provinces as Ilorin, Kabba and Benue. They are usually bought and sold as whole yams, but there is also a trade in yam flour and in dried slices. The yam may be eaten roasted whole, like a baked potato, or it may be pounded and the fibres partly removed. In this latter form it may be boiled; or frequently the pounded mass is put back into the outer shell of the yam and roasted.

CASSAVA (*MANIHOT* Sp.)

Cassava is universally grown throughout West Africa; but except in a few special areas it can hardly be regarded as one of the more important crops. There are two main varieties, namely sweet cassava (*Manihot Palmata*), and bitter cassava (*Manihot Utilissima*) which is poisonous when raw. Cassava is always grown from stem cuttings, and in the south after these are planted they receive no further cultivation. Here, cassava is always the last crop before cultivated land is allowed to revert to bush condition, and the crop is often not even weeded at all. In Northern Nigeria, however, it receives more attention, as it is one of the very few crops which will grow throughout the dry season. It may therefore be planted at the beginning of the rains and harvested throughout the dry season; or it may be planted towards the end of the rains and harvested during the early part of the following rainy season. In either case it is usually grown on specially made ridges, and is carefully weeded. The cassava fields moreover are usually surrounded by low fences during the dry season to keep goats, sheep and donkeys from damaging the crop.

In the coastal belts the yield of cassava varies enormously according to the length of time it remains in the ground; and also apparently according to the variety. On the Government farms in Southern Nigeria the ordinary yield of the commoner "bitter" variety is about 5 tons per acre when the crop has been in the ground for nine or ten months, but the crops of the ordinary native farmer are not nearly so good as those on the experimental farms. The "sweet" varieties yield about 3 tons per acre in Northern Nigeria, but apparently less in the south. Flour or pure starch is prepared from the cassava root by rather laborious processes of soaking or boiling, drying, grinding and sieving, and in Southern Nigeria the products form an important part of the diet of the people.

SWEET POTATOES (*IPOMOEA BATATAS*)

The sweet potato can hardly be regarded as one of the main crops of any part of West Africa; but it is very widely distributed and constitutes quite an important supplementary article of food, especially in the arid regions where yams are not usually grown. This crop is usually planted during the early rains; but even in the extreme north of Nigeria quite good yields are obtained if it is planted as late as early July; for it continues to grow quite late into the dry season. It does well as a dry weather crop under irrigation. Sweet potatoes will grow on almost any soil, but do best on one which is light, deep and well-drained. The plants are propagated by stem cuttings, which are planted on ridges or raised beds 18 inches to 2 feet apart. Weeding is necessary in the early stages, but the plant, being a creeper, when once established rapidly covers the ground. It is therefore very useful for preventing wash on exposed slopes. Grown as a rain crop an average yield is about 4 to 5 tons per acre, but much heavier crops can be obtained with irrigation and manuring.

COCO YAMS (*COLCOSIA ANTIQUORAM*)

Coco yams are largely grown in the cocoa belt of the Gold Coast. In Nigeria they are to be seen all over the country, but it is only in the extreme south-east and in the Cameroons that they are an important crop. In the north they are only grown on a very small scale in swampy land which is unsuitable for cereals. Their outstanding characteristic is that they will flourish in shade which would be too heavy for any other crop. They need ample moisture, and again will flourish where the rainfall is too high, or the ground too wet, for other crops.

In Northern Nigeria, coco yams are usually planted in May and harvested in November and December, but in the south they can be planted as late as July. The plants are propagated in the same way as yams; small tubers or the heads of large tubers are both used as "seed". They can be planted on the flat or on ridges. An average yield is about 3000 lb. of tubers per acre, but under favourable conditions 6000–7000 lb. per acre may be obtained.

BEANS

Practically every farmer in West Africa grows some type of bean, usually interplanted with other crops. The pigeon pea (*Cajanus indicas*) is restricted, for no very obvious reason, to some small, but widely scattered districts. The cultivation of the sword bean (*Cannavalia ensiformis*) seems, in Nigeria, to be restricted to some of the pagan tribes of the north. The cowpea (*Vigna Catjang*) of which there are many varieties not very different from each other, is grown throughout Nigeria. Lastly there are the innumerable, and very varied, types of local beans of the species *Phaseolus lunatus*. Some of the latter are grown in every district, though, in contrast to cowpeas, they take a more important place on the farms in the humid regions than in the arid ones. Excepting some of the cowpeas and *Phaseolus lunatus* types, these beans are all speckled or coloured, so that, although they are highly valued for food locally, they are unsuitable for export. Moreover, none of them forms a really dense cover to the land; and most of the varieties of *Phaseolus lunatus* suffer from the further disadvantage that they will give little yield unless they are provided with sticks or strings upon which to climb. The yield per acre of all these species of beans is relatively very small. 150 lb. per acre is a good yield when closely planted as a sole crop on the Government farms in Nigeria. On the ordinary farms, where they are thinly planted, and are often among other crops, the yield per acre must be very small indeed.

The right season for planting both cowpeas and beans in Nigeria is July or August, or even September in the south; for if they are planted earlier they produce little or no seed, however well they may grow. At this late period of the cropping season, the farmer in the coastal belt of all the colonies does not make full use of his land. He does indeed often plant some crop or other between the dying vines of his yams; but in many fields nothing is thus grown, and where anything is planted it is planted very thinly. The reason is that, with the exception of cotton, which cannot be sowed everywhere, none of the crops that can be grown at that time of the year gives any considerable yield per acre. Nor can the products of any of them, excepting only cotton, be sold readily wholesale. They have to be peddled in petty trade in the local markets, a method of disposal that clearly could not provide an outlet for an increased production.

It seems, therefore, that there should be an opportunity for the introduction of a bean suitable for export, especially as the price for dried beans of fair quality in Europe is normally from £25 to £35 per ton, which is sufficient to leave a fair price for the producer in West Africa after all expenses have been deducted.

An increase in the cultivation of beans would be advantageous in view of the effect which they would have in maintaining the soil in a state of high fertility under permanent cultivation (see Chapter 6 on green manuring). It is hardly to be expected that a productive bean could be found that would be as efficient in smothering weeds and improving the soil as a plant such as *Mucuna*, which is selected specially for these two purposes. But a bean might well be found that would be of some value for these purposes and also be suitable for export. An effort has therefore been made recently by the Agricultural Department in Nigeria to find such a bean. Many varieties have been introduced for trial from several different countries, but without success. More even than in other crops, introduced varieties of beans seem to be greatly damaged by insects, which either are not attracted to the local varieties at all, or do them but little damage. It seems likely, however, that a suitable bean has been produced by selection and breeding; but it will take some years to multiply it on a large scale, so that it can be grown in commercial quantities.

THE BAMBARA GROUNDNUT
(*VOANDIZEIA SUBTERRANEA*)

The Bambara groundnut is widely cultivated in West Africa for local consumption, but has no commercial value for export.

IRRIGATED CROPS

There are no great irrigation schemes in West Africa, but the native farmer of Northern Nigeria is by no means ignorant of the value of irrigation. In almost every river valley in the more arid regions there are small fenced patches of land on to which water is lifted; and they are very intensively cultivated. At Sokoto there is a small experimental irrigation scheme which is financed by the native administration and supervised by the Agricultural Department.

The chief crop grown in the irrigated gardens is

onions; but tomatoes are becoming common, and excellent tomatoes can now be purchased in almost any market of fair size in Northern Nigeria. The onion is a very paying crop; it yields well and fetches high prices. In Sokoto and the extreme north generally wheat is also grown as a dry season crop by means of irrigation. It is, however, a luxury foodstuff and there is no general demand for it.

THE CROPS OF THE "COMPOUND"

Around almost every compound there is a small area of land which is heavily manured by the refuse from the compound, and this is usually used for growing minor crops for the domestic needs of the household. Among these may be mentioned okra (*Hibiscus esculentus*), which is a favourite vegetable, rama (*Hibiscus Cannabinus*), grown for fibre, gourds, dye plants such as henna and indigo, "red peppers" and tobacco.

Okra and rama are also often grown on a field scale in the south, generally interplanted with yams. Okra is cultivated for its fruits, but the leaves are also eaten. Both indigo and henna are also sometimes grown on a field scale, although one rarely sees plots of any size. Gourds and pumpkins of many and very varied types and uses are grown through West Africa, both in the fields, where they are thinly scattered among other crops, and in the compounds, where they straggle over the fences or odd corners of ground.

CHAPTER 14

LIVE STOCK

THE domestic live stock of West Africa includes horses, donkeys, cattle, pigs, sheep, goats, fowls, guinea-fowl, turkeys and guinea-pigs. Cattle and horses are practically confined to Northern Nigeria and the Northern Territories of the Gold Coast. Sheep, goats and fowls are universal, but pigs are only found in non-Muhammedan areas.

Nearly every farmer owns a few head of sheep or goats, and in many parts of the south they form almost the only source of meat for the ordinary people. There are several distinct types of both sheep and goats; those of the south are very small with comparatively short legs, while those of the north are large and leggy. Some of the Hausa goats are very big indeed and appear to be good milkers. In British West Africa, all the types of sheep are hairy, not woolly; but in the French Sudan there is a breed of woolly sheep, and experiments are being conducted by the Nigerian Veterinary Department to test the possibility of breeding a race of wool-bearing sheep by crossing the local sheep with imported types. Apart from this, no serious attempt has ever been made to improve West African sheep or goats whether from the point of view of meat, milk or skins; nor has their value as manure makers ever been exploited by folding. They roam freely over the fields during the day in the dry season, picking up whatever fodder they can find; during the rains they are kept in the compound by day as well as night in order to prevent them from damaging crops. The skins of both sheep and goats are a valuable article of commerce, and those from parts of Northern Nigeria are of very high quality; but many of them, perhaps 60 per cent., are spoilt by disease and bad flaying. The Veterinary Department has in the last few years, with considerable success, made efforts to improve the flaying, especially in the bigger towns.

In Nigeria an attempt has been made to start a flock of sheep on the experimental farm at Ilorin; but this attempt has been very severely handicapped by heavy losses due to disease, especially among some sheep which were taken down from farther north. Very little is known about the diseases of sheep and goats; but most of the losses at Ilorin have been due to worms and lung trouble. It seems to be characteristic of all stock in West Africa that they have become strictly acclimatized to the particular conditions of the district in which they are normally found; and when they are moved to an area where the conditions are even slightly different, very heavy losses occur before they become re-acclimatized to their new conditions.

As with sheep and goats, there are several different types of cattle. These types are so distinct that they might well be called "breeds". Some of the big Fulani herds are nearly pure to type as regards colour and minor points, and the Fulani appear definitely to try to maintain the type by the selection of bulls.

In the south there is a breed of dwarf cattle, which have little economic value, but are kept almost entirely for ceremonial purposes or for dowries. They are generally to be seen in rather small herds; but there are a few head in most villages. They are very docile, and are allowed to roam, more or less at large, in the surrounding bush in search of food, but come, or are brought back, to the village at night. They frequently do considerable damage to crops. The interesting point about these small cattle is the fact that, although they are apparently not entirely immune to trypanosomiasis, they have developed a very considerable degree of resistance and can exist in areas where, owing to the prevalence of the tsetse fly, the bigger animals of the north would die in a few weeks.

The cattle of the north can be divided into long-horned, and short-horned breeds. The long-horned are associated with the nomadic mode of living, while the short-horned types are, in Nigeria, to be found chiefly in Bornu and Sokoto where they are kept under more settled conditions. The Sokoto cattle are very similar in appearance to the "Montgomery" or "Sanniwal" breed of India. They have big frames and a heavy dewlap; many of them have the general appearance of potential milkers. The prevailing colours are white and a peculiar, but striking, dun or silver-grey. The Bornu cattle are smaller than the Sokoto cattle and are mostly red or black. They are also very docile. Among the truly nomadic cattle there are at least two well-marked types. Of these one is a white breed with black points and the other is entirely red and has particularly long horns. There are of course many intermediates, and crosses between all these types; and many of the cattle of Northern Nigeria do not fit exactly into any one of the main types. All the local cattle, of whatever breed, are slow growing, and, apart from a few exceptional individuals, the cows give very little milk. They also tend to be "narrow" or "flat sided", and rather leggy.

In view of the importance of the cattle industry and the possibility of introducing mixed farming, the improvement of the local cattle by selection has been undertaken in Northern Nigeria, and a Government stock farm has been started near Zaria solely for this purpose. On this farm an attempt is being made to improve, by selective breeding, the milking qualities of the cows; and simultaneously to study the practical problems of the maintenance and feeding of cattle in the conditions obtaining on local native farms. Under nomadic conditions, the cattle have abundance of food and water during the rainy season; but during the dry season the grazing becomes exceedingly poor, and the cattle, receiving no additional food, then pass through a period of two or three months of partial starvation, when, living chiefly on the food reserves in their bodies, they lose weight and condition at a rapid rate.

Among the Fulani herdsmen the selection of sires is practised to some extent, and this serves to maintain the type; but all the females are used for breeding, and the great object of the herdsman is to increase the size of his herd. For in the past the losses from epidemics, aided by the weakening effects of the annual period of starvation, caused such great losses that numbers were inevitably the first consideration.

The Fulani has evolved an animal which is adapted, by very slow growth and infrequent breeding, to the conditions under which it lives; and it seems probable that if all the cattle had always to live in these conditions little improvement would be possible, beyond the gradual elimination of the poorer stock, as veterinary measures gradually lessened the necessity to maintain excessive numbers as an insurance against the next epidemic. A fast-growing animal would not survive in these conditions; cows that were capable of breeding frequently and giving much milk would either be unable to fulfil their capabilities or would die of starvation.

Experience on the Government farms has already shown that, once the annual period of starvation is eliminated, growth becomes more rapid, the animals mature at an earlier stage, and begin to breed more regularly. The milk yield also is improved even among unselected animals; it is found that heifers that have never been starved all give more milk than their mothers who had thus suffered annually during their early years.

It is also quite clear that there are very great possibilities in selective breeding for milk; for there are among the cows quite distinct milking and non-milking strains. The former will respond to good feeding and increase their yield, but the latter will always remain poor milkers no matter how well they are fed. Whether there is any other factor, beyond the food supply, which tends to limit or retard improvement remains to be seen; but it seems unlikely. The cattle, even under nomadic conditions, show no signs of specific mineral deficiency; and the experimental feeding of bone meal has given negative results, whether given in combination with concentrates or to cattle who have to live on grazing alone. It is therefore confidently anticipated that by selective breeding, combined always with adequate feeding, it will be possible very quickly to raise the level of milk production from the 80–100 gallons per lactation, which is about the average at present, to the neighbourhood of 300 gallons per lactation, and simultaneously to increase the rate of breeding, so that even good milkers will produce a calf every year. Under nomadic conditions the good milkers commonly produce a calf only once in every two years. This is no doubt due to the fact that an animal, under such conditions, cannot bear the strain of milk production and reproduction at the same time.

Mixed farming is therefore one of the essentials for continued improvement of the cattle, and experience already shows that once the settled farmer takes to the keeping of cattle, he will devote adequate attention to their feeding in the dry weather. It pays him to do so, and moreover the native of Northern Nigeria seems commonly to have an instinctive love of cattle, and to take a pride in the condition of his beasts. As milk is one of the most valuable commodities which can be produced on the farm, it is reasonable to assume that dairying will be a paying proposition, even if it involves the feeding of grain. But it is as yet doubtful whether the feeding of grain will be necessary until the average yield of the cows is higher than it is at present. Some of the by-products of the farm—groundnut leaves, cowpea leaves, and guinea corn leaves—make excellent "long fodder", which will at least maintain condition, while guinea corn bran is a most valuable form of "concentrate" for cattle feeding, and is at present so little valued that it can be bought for next to nothing from any farmer. It is difficult to over-estimate the importance of increasing the milk supply of the country, either from the nutritional or the economic point of view; and there seems to be no reason why a considerable degree of improvement should not be attained.

The Nigerian Government has decided to concentrate on improvement of milk production; but value for work in the males is regarded as the secondary object of the breeding programme. Most of the male cattle can be easily trained to work; but, as might be expected, some work much better than others. This seems to be a matter of temperament as much as of build; but whether temperament is an individual characteristic or a breed characteristic has not yet been ascertained. At present three different breeds are being tested, and the one which best combines heavy milking cows with good working bulls will eventually be the one which will be concentrated upon.

No serious attempt has ever been made to improve the cattle of West Africa by importing improved cattle from other parts of the world. A few animals have very occasionally been imported; but they have never lived very long. With promising material already in the country, strains which have developed some degree of resistance to local diseases, it would seem unwise and even dangerous to introduce other less-resistant breeds from outside.

The chief epidemic diseases of the local cattle, rinderpest, pleuro-pneumonia and black quarter, can now be controlled and prevented by the Veterinary Department; were this not the case there would be little hope for any improvement in the stock or local methods of stock-keeping. Apart from these three diseases, trypanosomiasis is the most serious cause of mortality among the cattle, but this can now be effectively treated;

and, moreover, it is hoped that gradually those fly belts, which, owing to their position cause most trouble, will be cleared.

Under settled conditions, our own experience has shown that a skin disease—streptothricosis—can be very troublesome, and may cause serious losses; but this can be both prevented and cured by dipping the cattle in arsenical dips. It is possible that this disease may not generally give so much trouble to native mixed farmers as it does on the relatively big Government farms. If it is found to do so, it will be necessary to build dips to which native farmers can bring their cattle.

Apart from these diseases, the cattle appear to be remarkably hardy and healthy. The Fulani herdsmen regularly supply their cattle with either a little salt or "kanwa" (one of the local salts); and they have their own remedies for minor ailments, which appear to be more or less effective. Calf mortality is, however, higher than it should be, both on the experimental farm and among the Fulani, and requires thorough investigation.

There is considerable export trade in hides from Northern Nigeria, the skins being obtained from the large numbers of cattle which are annually slaughtered for beef. The possibility of making use of the bones and blood from the slaughter-houses for manurial purposes has from time to time been considered, but so far has not been found to be a practical proposition.

Horses are of no great agricultural importance. They are, however, bred in considerable numbers in the north, and a breeding mare is a great asset to a farmer. When well fed and cared for under European supervision the local horses become useful hacks and polo ponies: it is possible that, like the cattle, they could be considerably improved by better feeding while young and by selective breeding. The Emirs of the north have recently evinced considerable interest in the improvement of their horses, and Arab stallions have been imported into Bornu with this end in view. Previous attempts to import English sires have not been a success. The effect of better feeding of mares and foals does not seem to have been tried; but a certain amount of selection of sires is undoubtedly practised.

Donkeys are rarely seen in the coastal belts, but they are used as pack animals in Northern Nigeria in large numbers. The West African fowl is one of the jokes of the coast, but that it plays an important part in native economics is shown by the fact that fowls are to be found in almost every compound, and are bought and sold in every market. They can be greatly improved by crossing with English poultry. The cross-bred bird seems to be quite as hardy as the native one, and lays a very good-sized egg. The Nigerian Agricultural Department has for some years kept a small number of Rhode Island Red fowls at Ibadan, and sold eggs for setting to local natives. This is now beginning to have a very noticeable effect on the fowls of that district.

INDEX

Millet (gero), 25, 38, 39, 140–142
Mixed farming, 48, 64
Mucuna, 53, 55, 56
Munshis, 24, 127, 128, 129

Native chiefs and extension work,
85–87
Native farmer, and cattle, 65
and new industries, 6, 32, 76,
83, 84, 90, 128
and world trade, 27–30
methods of, 6–12
Native labour, 45, 46
Niger Delta, the, 19, 44, 48, 105
Nigeria, 13, 15–17, 109, 112, 113,
134, 151
Co-operative Societies in, 89
land in, 43
Northern, 15, 18, 20, 24, 33,
118, 120, 123, 127, 136, 137;
drought in, 38, 142, 143, 156,
159; farming methods in, 37
Southern, 20, 24, 105, 115, 135

Oil palm, the, 19, 43
age of fruiting, 97
and annual crops, 41
and rainfall, 93
breeding of, 100, 101
distribution of, 93
harvesting of, 27
methods of growing, 96–99
ownership of, 26, 27, 96
Oil press, 102–103
Onions, 155
Onitsha, 44
Owerri, 44

"Palm Grove Improvement",
98–100
Palm oil, and kernels, 104
and margarine making, 103
expression of, 102, 103
extraction of, 101
"hard" and "soft", 102–104
Peasant proprietorship, 31, 33

Peasants and new industries, 6, 32,
76, 83, 84, 90, 128
Permanent crops, 39–42
Phosphate, use of, 51, 52
Planting companies versus peasant
proprietorship, 31–33
Plough, and mixed farming, 66–69
types of, 67, 68
Potash, use of, 51, 52

Rainfall, 13–17
and cocoa, 105
and oil palms, 93, 105
Rice, 142–145
Rights in land, 26, 27
Rinderpest, 65
Rotation of crops, 36, 46, 47, 60,
62

Season, dry, 13–17, 37, 105
variations in work in, 11, 37
wet, 13–17, 37
Secondi, 14
Shea nut, 27
Sheep, 156, 157
use on farms, 74, 75
Shifting cultivation, 25, 38, 39,
43 et seq.
Sierra Leone, 15, 28, 93, 95, 114,
118, 131, 142
Soil, acidity of, 48, 49, 105
biological changes in, 20
chemical composition of, 18–20
differences in, 18
minerals in, 51
nitrogen in, 50–63
relation of, to livestock, 20
Sokoto, 14, 16, 158
Streptothricosis, 74, 163
Sweet potatoes, 150, 151

Taxation, direct, 29
Temperature, 13, 14, 17
Tenure of land, 21–23, 39, 40,
42
Tomatoes, 155